AF735212

A Practical Guide to Implementing Statistical Methods

First Edition

Aina Katsikas

University of Nevada, Reno

cognella®
SAN DIEGO

Bassim Hamadeh, CEO and Publisher
Christina Brown, Acquisitions Editor
Skyler Bryant, Project Editor
Samantha Hansen, Production Editor
Gwendolyn Teeling, Editorial Associate
Jessica Delia, Graphic Design Associate
Alexa Lucido, Senior Licensing Manager
Tiffany Mok, Senior Content Marketing Specialist
Natalie Piccotti, Director of Marketing
Kassie Graves, Senior Vice President of Editorial
Alia Bales, Director of Project Editorial & Production

Copyright © 2027 by Cognella, Inc. All rights reserved. No part of this publication may be reprinted, reproduced, transmitted, or utilized in any form or by any electronic, mechanical, or other means, now known or hereafter invented, including photocopying, microfilming, and recording, or in any information retrieval system without the written permission of Cognella, Inc. For inquiries regarding permissions, translations, foreign rights, audio rights, and any other forms of reproduction, please contact the Cognella Licensing Department at rights@cognella.com.

Trademark Notice: Product or corporate names may be trademarks or registered trademarks and are used only for identification and explanation without intent to infringe.

Cover image copyright © 2024 iStockphoto LP/MicroStockHub.

Printed in the United States of America.

cognella® ACADEMIC PUBLISHING
320 South Cedros Ave., Ste. 400, Solana Beach, CA 92075

Contents

1 Data Analytics Part 1

1.1 Qualitative Data

Qualitative data describes attributes or characteristics, such as hair color. Unlike quantitative data, which represents numerical values like weight or height, qualitative data categorizes observations based on non-numerical traits. Any information that can be measured or assigned a numerical value falls under quantitative data.

Qualitative data typically consists of **categorical variables**, which contain multiple categories under a broader qualitative characteristic. For example, the categorical variable **hair color** includes categories such as brown, black, and blonde. Each color represents a distinct category within this variable.

Visualizing Qualitative Data

The two most common ways to visualize qualitative data are **bar charts** and **pie charts**.

Bar Charts

A bar chart displays the **frequency distribution** of qualitative data, showing how often each category appears. A **frequency distribution** lists the frequency (or relative frequency) of each category within a categorical variable.

For example, if we collect data on the hair color of students in a course, we can create a frequency distribution to summarize the number of students with each hair color. A bar chart would then provide a visual representation of this distribution.

Steps to Create a Bar Chart

1. **Collect Data** – Use the dataset provided in the Excel resource *LectureData.xlsx* (under the "Pie and Bar Charts" tab).

2. **Calculate Relative Frequency** – Divide the frequency of each category by the total number of observations. Suppose we surveyed 100 students and found:

   - 31 students with brown hair
   - 32 students with black hair
   - 37 students with blonde hair

   The relative frequencies are:

   - Brown: 0.31 $(31/100)$
   - Black: 0.32 $(32/100)$
   - Blonde: 0.37 $(37/100)$

3. **Create the Bar Chart** – In Excel:

   - Select **cells A3:A5**
   - Hold **Command** and select **cells C3:C5**

- Click **Insert > Bar Chart** and choose the first option.

The bar chart will display the relative frequencies.

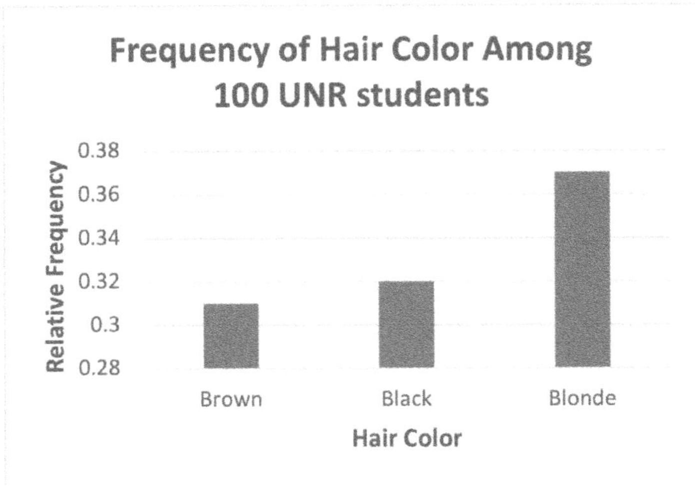

Image 1.1: Frequency of Hair Color Among 100 UNR Students

Bar charts can also display raw frequency instead of relative frequency. If you prefer to visualize the number of students in each category, select **Column B instead of Column C** when creating the chart.

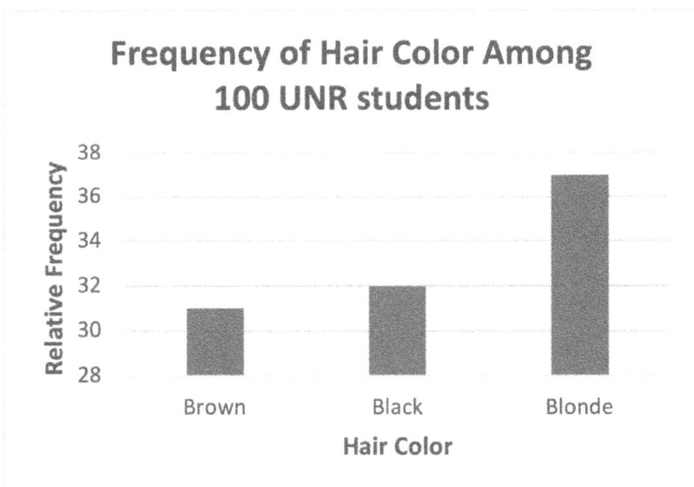

Image 1.2: Frequency of Hair Color Among 100 UNR Students

Pie Charts

A **pie chart** is another useful way to present qualitative data. It is a circular graph where each segment represents the **relative frequency** of a category within a categorical variable.

Using the same hair color data, follow these steps to create a pie chart in Excel:

1. Select **cells A3:A5**

2. Hold **Command** and select **cells C3:C5**
3. Click **Insert > Pie Chart** and choose the first option

This will generate a pie chart that visually represents the proportion of students with each hair color.

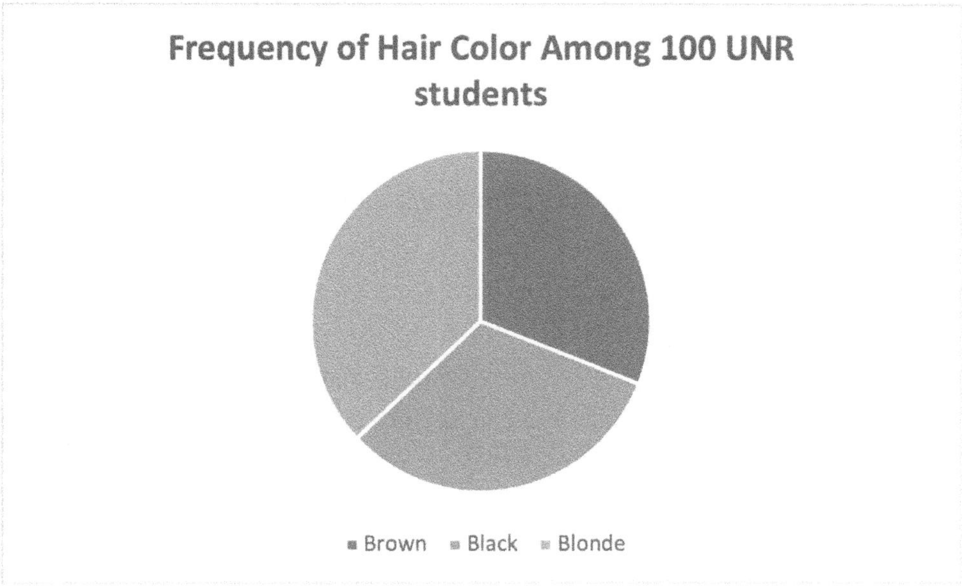

Image 1.3: Frequency of Hair Color Among 100 UNR Students

1.2 Quantitative Data

Quantitative data consists of numerical values, including integers and decimals. Qualitative data answers the question "what type" whereas quantitative data answers the question "how much".

Histograms

A histogram is a popular method for visualizing quantitative data. It consists of a series of adjacent rectangles, where:

- The **width** of each rectangle represents the interval width.
- The **height** of each rectangle represents the frequency (or relative frequency) of data points within that interval.

Creating a Histogram in Excel

To create a histogram, use the "*Histograms*" tab in the *LectureData.xlsx* file. This dataset contains sales transaction data from 41 transactions at an autoshop in Reno, Nevada. For example, one transaction costs $150, while another costs $101.

Follow these steps in Excel:

1. Select **Data > Data Analysis > Histogram > OK**.
2. If Data Analysis is not available, enable the **Analysis ToolPak** add-in.
3. Set the parameters as follows:
   - **Input range:** Select cells `$A$1:$A$41`.
   - **Bin range:** Select cells `$C$1:$C$7`.
   - Check the **Labels** box.
   - **Output range:** Select `E1`.
   - Check the **Chart Output** box.
4. Click **OK**.

This will generate a histogram similar to the one shown below:

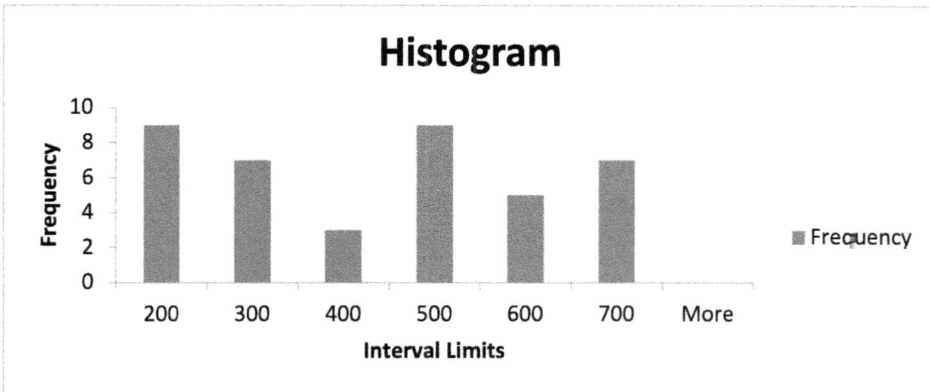

Image 1.4: Frequency of Hair Color Among 100 UNR Students

Each interval on the x-axis represents a range of sales transaction prices, while the height of each bar indicates the frequency of transactions within that range. For example:

- There are nine transactions in the \$100 - \$200 range.
- There are seven transactions in the \$201 - \$300 range.
- The "More" category represents transactions beyond the highest defined interval, which in this case is 0.

This histogram provides a clear visualization of the distribution of transaction amounts.

1.3 Mean, Median, and Mode

Measures of Central Location

A measure of central location identifies a typical or central value that represents the dataset. The three primary measures of central location are the **mean**, **median**, and **mode**. Each measure provides a different perspective on the data distribution.

Mean

The arithmetic mean is the most commonly used measure of central location. It is calculated by summing all observations and dividing by the total number of observations.

For example, consider a dataset representing the ages of six students in Reno, Nevada:

$$\text{Mean} = \frac{12 + 13 + 15 + 14 + 11 + 10}{6} = 12.5$$

Median

Although the mean is widely used, it is highly sensitive to outliers. These are values that are significantly higher or lower than the rest of the dataset. For example, if the last student in the data above was 65 years old (rather than 10 years old), the mean would be 21.6. This mean is less accurate of a representation compared to 12.5.

The median, on the other hand, is the middle value in an ordered dataset. Half of the observations lie above the median, and half lie below. If the dataset contains an even number of observations, the median is the average of the two middle values.

To calculate the median, first arrange the data in ascending order:

$$10, 11, 12, 13, 14, 15$$

Next, eliminate the smallest and largest values, moving inward:

12, 13

The dataset has an even number of observations, therefore the median is the average of the two middle values:

$$\text{Median} = \frac{12 + 13}{2} = 12.5$$

Unlike the mean, the median is resistant to extreme values, making it a more reliable measure of central tendency for skewed data. In this case, the mean and median are equal, indicating a symmetric (normally distributed) dataset.

Mode

The mode is the most frequently occurring value in a dataset. A dataset may have no mode, one mode, or multiple modes.

Using the original student age dataset:

$$12, 13, 15, 14, 11, 10$$

Since no value appears more than once, this dataset has no mode.

Now, consider a different dataset:

$$13, 13, 15, 15, 11, 13$$

Here, the number 13 appears three times, more frequently than any other value. Thus, the mode of this dataset is:

$$\text{Mode} = 13$$

Calculating Measures of Central Location in Excel

Excel provides built-in functions to compute these measures:

- To calculate the mean: `=AVERAGE(array)`
- To calculate the median: `=MEDIAN(array)`
- To calculate the mode: `=MODE(array)`

Replace `array` with the selected data range in Excel.

1.4 Percentiles

The median is a measure of central location that divides a dataset in half and is also known as the fiftieth percentile. A box plot is a useful visual tool for representing different percentiles in a dataset.

In general, the **Pth percentile** divides a dataset into two parts:

1. Approximately **P%** of the observations have values **less than** the Pth percentile.
2. Approximately **(100 minus P)%** of the observations have values **greater than** the Pth percentile.

For example, if a student lands in the eightieth percentile of SAT scores, their score is higher than 80% of other test-takers.

We will use **Excel** to calculate percentiles for **sales transactions** at an auto shop in Reno. Using the **Percentiles** tab in the **LectureData.xlsx** file, we can first determine the minimum and maximum values using the following formulas:

- `=MIN(A2:A41)` to find the **minimum value** (101)
- `=MAX(A2:A41)` to find the **maximum value** (681)

Next, we calculate key percentiles:

- `=PERCENTILE.INC(A2:A41, 0.25)` to find the **twenty-fifth percentile**
- `=PERCENTILE.INC(A2:A41, 0.50)` to find the **fiftieth percentile (median)**
- `=PERCENTILE.INC(A2:A41, 0.75)` to find the **seventy-fifth percentile**

From this, we see that:

- 25% of the sales transactions are **less than 226.25**
- 50% of the sales transactions are **less than 426.5**
- 75% of the sales transactions are **less than 536**
- 100% of the sales transactions are **less than 681**

Min	Twenty-fifth percentile	Fiftieth percentile	Seventy-fifth percentile	Max
101	226.25	426.5	536	681

1.5 Range, Variance, and Standard Deviation

In statistics, three key measures describe the spread, or dispersion, of data: **range, variance, and standard deviation.** Dispersion helps us understand how much the data varies from the mean. While the mean provides a central value, it does not always tell the full story. Consider the following two datasets, each representing quiz scores from different classes:

	Class A	Class B
	2	3
	2	3
	2	1
	2	1
Mean	2	2

Both classes have the same mean quiz score of 2, but the distribution of scores is quite different. This is why measures of dispersion are crucial—they provide insight into how spread out the data is.

Range

The **range** is the simplest measure of dispersion, calculated as the difference between the maximum and minimum values in a dataset.

For Class A:

$$\text{Range} = 2 - 2 = 0$$

For Class B:

$$\text{Range} = 3 - 1 = 2$$

The range shows that Class A's scores are uniform (all students scored the same), while Class B's scores vary. However, range alone does not provide a complete picture of dispersion since it only considers the extremes.

Variance and Standard Deviation

Variance and standard deviation are more informative measures of dispersion. They measure the average distance of the values from the mean. The standard deviation offers a metric that's easier to interpret. The **variance** quantifies the average squared deviation from the mean, while the **standard deviation** is the square root of the variance.

The formulas for standard deviation and variance are:

- **Sample Standard Deviation**:

$$s = \sqrt{\frac{\sum(x_i - \bar{x})^2}{n-1}}$$

- **Sample Variance**:

$$s^2 = \frac{\sum(x_i - \bar{x})^2}{n-1}$$

- **Population Standard Deviation**:

$$\sigma = \sqrt{\frac{\sum(x_i - \mu)^2}{N}}$$

- **Population Variance**:

$$\sigma^2 = \frac{\sum(x_i - \mu)^2}{N}$$

where:

- x_i represents each data point,
- $\bar{x}$ (sample mean) or μ (population mean) represents the center of the data,
- n is the sample size, and N is the population size.

Calculating Standard Deviation and Variance

Let us compute the **sample** standard deviation and variance for Class A and Class B.

Class A Since all scores are identical (2), each deviation from the mean is zero:

x_i	$\bar{x}$	$x_i - \bar{x}$	$(x_i - \bar{x})^2$
2	2	0	0
2	2	0	0
2	2	0	0
2	2	0	0
$\sum$			0

$$s = \sqrt{\frac{0}{n-1}} = \sqrt{\frac{0}{3}}$$

$$s = \sqrt{0} = 0$$

Both the variance and standard deviation are **0**, meaning there is no spread—the data points are identical.

Class B Now, let us compute the standard deviation for Class B.

x_i	$\bar{x}$	$x_i - \bar{x}$	$(x_i - \bar{x})^2$
3	2	1	1
3	2	1	1
1	2	-1	1
1	2	-1	1
$\sum$			4

$$s = \sqrt{\frac{4}{3}} = \sqrt{1.3333}$$

$$s = \sqrt{1.3333} \approx 1.1547$$

The standard deviation of **1.1547** indicates that the scores in Class B are more spread out than those in Class A.

Key Takeaways

- The **range** provides a quick measure of spread but only considers the maximum and minimum values.
- The **variance** accounts for all data points by computing squared deviations from the mean.
- The **standard deviation** gives a more intuitive measure of spread, in the same units as the data.

By considering these measures, we can better understand how much variation exists within a dataset, rather than exclusively relying on the mean.

2 Data Analytics Part 2

2.1 Sample versus Population

In statistics, it is important to distinguish between a **population** and a **sample**.

- A **population** includes *all* individuals or observations of interest in a study. It represents the entire group from which data could be collected.
- A **sample** is a *subset* of the population, selected to represent the larger group. Since gathering data from an entire population is often impractical, researchers typically analyze a sample to make inferences about the population.

Example: Suppose a university wants to determine the average GPA of all students.

- If the university calculates the average GPA using data from **all students**, then the data represents the **population**.
- If the university selects **200 students** at random and calculates their average GPA, this is a **sample**, which is used to estimate the average GPA for the entire student body.

Samples provide an estimate of the population. Statisticians use methods such as hypothesis testing to measure the reliability of their conclusions. Choosing a sample that is representative of the population is crucial for obtaining accurate results. A helpful analogy is to think of the sample as a slice from the pizza, which is the population!

2.2 Measures of Association

There are two key measures of association that quantify the relationship between variables: covariance and correlation.

The covariance measures the direction of the linear relationship between two variables (i.e., whether the relationship is positive or negative). There are two formulas to calculate the covariance depending on whether you are calculating the sample or population covariance.

Sample Covariance:

$$s_{xy} = \frac{\sum(x_i - \bar{x})(y_i - \bar{y})}{n - 1}$$

- x_i is each data point of the x variable,
- $\bar{x}$ is the mean of the x variable,
- y_i is each data point of the y variable,
- $\bar{y}$ is the mean of the y variable,
- n is the sample size

Population Covariance:

$$\sigma_{xy} = \frac{\sum(x_i - \mu_x)(y_i - \mu_y)}{N}$$

- x_i is each data point of the x variable,
- μ_x is the mean of the x variable,
- y_i is each data point of the y variable,
- μ_y is the mean of the y variable,
- N is the population size

However, in this textbook, we will learn to calculate the covariance using Excel.

Using the "Measures of Association" tab in the `LectureData.xlsx` file, we will calculate the covariance between income and education for twenty individuals. In any blank cell, type the command:

$$= \text{COVARIANCE.S}(A2 : A21, B2 : B21)$$

The resulting covariance is 31.4737. Covariance can take any value, including negative, positive, or zero. The interpretation of covariance is as follows:

- A negative value indicates a negative linear relationship between the two variables. That is, when x is above its mean, y tends to be below its mean, and vice versa.
- A positive value indicates a positive linear relationship. When x is above its mean, y tends to be above its mean, and vice versa.

- A covariance of zero indicates that the two variables do not have a linear relationship.

In this case, the covariance is positive, suggesting that income and education have a positive linear relationship.

The correlation, on the other hand, measures both the direction and the strength of the linear relationship between two variables. While covariance provides the direction of the relationship, correlation provides an indication of the strength of the relationship. You can think of covariance as indicating the direction your car is driving (north or south), whereas correlation measures both the direction and the speed (e.g., north or south, fast or slow).

There are two formulas to calculate the correlation coefficient, depending on whether you're calculating the sample or population correlation.

Sample Correlation:

$$r_{xy} = \frac{s_{xy}}{s_x s_y}$$

where s_{xy} is the covariance, s_x is the standard deviation of the x variable, and s_y is the standard deviation of the y variable.

Population Correlation:

$$\rho_{xy} = \frac{\sigma_{xy}}{\sigma_x \sigma_y}$$

where σ_{xy} is the population covariance, and σ_x and σ_y are the population standard deviations of the x and y variables, respectively.

In this textbook, we will learn to calculate and the correlation using Excel.

Using the same file, we will calculate the correlation between income and education for twenty individuals. In any blank cell, type the command:

$$= \text{CORREL}(A2 : A21, B2 : B21)$$

The resulting correlation coefficient is 0.5789, which indicates a positive linear relationship. The correlation coefficient always lies between -1 and 1, and its interpretation is as follows:

- A correlation of 1 indicates a perfect positive linear relationship.
- A correlation of -1 indicates a perfect negative linear relationship.
- A correlation greater than 0.5 indicates a strong positive linear relationship.
- A correlation between 0 and 0.5 indicates a weak positive linear relationship.
- A correlation less than -0.5 indicates a strong negative linear relationship.
- A correlation between 0 and -0.5 indicates a weak negative linear relationship.

Given that the correlation between income and education is 0.5789, we can conclude that there is a strong positive relationship between these two variables. This makes sense because, generally, an increase in the number of years of education leads to a higher income.

3 The Basics of Hypothesis Testing

3.1 Introduction to Hypothesis Testing

We make decisions based on our beliefs, holding hypotheses about various situations, which guide our choices. For example, I may hypothesize that the average height of adult women in the United States is 5'4". Eventually, we'll test these hypotheses against evidence, and then will be able to reject or not reject them. In this case, evidence would come from measuring the heights of a random sample of adult women in the United States to see if the average height is indeed 5'4".

For every hypothesis, there is an alternative contradicting scenario. In this case, the alternative hypothesis would be that the average height of adult women in the United States is not 5'4". **The current perspective or the status quo is the null hypothesis, and the contradictory scenario is the alternative hypothesis. Our goal is to determine if the null hypothesis is false and should be rejected. We always begin a hypothesis test by assuming the null is true.**

The null hypothesis is H_0: The average height of adult women in the U.S. is 5'4", and the alternative is written as H_A: The average height of adult women in the U.S. is not 5'4".

If the evidence contradicts the null hypothesis, we reject it. For example, if evidence suggests that the average height of adult women is not 5'4", we reject the null. However, if the evidence is consistent with the null, we do not reject it. This is a subtle distinction because we are not accepting the null hypothesis. Not rejecting the null is NOT the same as verifying it's truth. By stating that we do not reject the null, we mean that it is possible, but not certain, that the average height is 5'4". This is similar to a court case: we may have enough evidence to prove someone is not innocent but not enough to prove they are guilty. Similarly, we cannot conclusively prove that the average height of adult women is 5'4", only that it might be.

There are rules on how to structure hypothesis statements.

We need to avoid vague hypotheses. For example, the hypothesis "The average height of adult women in the United States is short" is too broad. It must be specific:

$$H_0 : \text{The average height of adult women in the U.S. is 5'4"}.$$

The corresponding alternative is:

$$H_A : \text{The average height of adult women in the U.S. is not 5'4"}.$$

This is an example of a two-tailed test. We reject the null hypothesis (that the average height is 5'4") if we find that the actual average height is either greater than or less than 5'4". The test is called "two-tailed" because we can reject the null hypothesis on either side of the hypothesized value.

Alternatively, we can set up a one-tailed test:

$$H_0 : \text{The average height of adult women in the U.S. is} \leq \text{5'4"}.$$

The corresponding alternative is:

$$H_A : \text{The average height of adult women in the U.S. is} > \text{5'4"}.$$

In this case, we can reject the null hypothesis only if the actual average height exceeds 5'4". This is a right-tailed test because rejection occurs only on the right side of the hypothesized value.

An easy trick to distinguish between right-tailed, left-tailed, and two-tailed hypotheses involves the mathematical sign used in the alternative hypothesis. The table below summarizes this concept:

Type	Sign in alternative	Null	Alternative
Right-Tailed	$\geq, >$	The average height is < 5'4".	The average height is $\geq$ 5'4".
Left-Tailed	$\leq, <$	The average height is > 5'4".	The average height is $\leq$ 5'4".
Two-Tailed	$=, \neq$	The average height is = 5'4".	The average height is $\neq$ 5'4".

The Difference Between the Null and Alternative Hypotheses

The null hypothesis represents the status quo, what we currently believe to be true. The alternative hypothesis challenges the null, suggesting a different scenario. The alternative is always the opposite of the null and mutually exclusive from the null. For example, the following pair of hypotheses is incorrect:

$$H_0 : \text{The average height is } \leq \text{5'4".}$$

$$H_A : \text{The average height is } \geq \text{5'4".}$$

These hypotheses overlap, and the evidence could fall within this overlap (e.g., if the average height is exactly 5'4". A better formulation would be:

$$H_0 : \text{The average height is } \leq \text{5'4".}$$

$$H_A : \text{The average height is } > \text{5'4".}$$

How to Determine the Alternative Hypothesis in a Word Problem

In a word problem, the alternative hypothesis is often indicated by phrases like "test if", "test the claim", or "wants to determine if".

Correct Notation (Mean versus Proportion)

We must also consider whether to use the μ symbol (population mean) or p (population proportion). For example, if we are testing the mean, we use μ (pronounced mew):

$$H_0 : \mu = 5'4"$$

$$H_A : \mu \neq 5'4"$$

If we are testing the proportion, we use p, see example 2 below for a demonstration.

Examples

Example 1:

A consulting firm hypothesizes that the average value of back-to-school spending per family will be \$600 this year. They want to test if the average spending will differ from this amount. The null and alternative hypotheses are:

$$H_0 : \mu = 600$$
$$H_A : \mu \neq 600$$

Example 2:

A television analyst wants to test if more than 50% of households will tune in for a TV episode. The null and alternative hypotheses are:

$$H_0 : p \leq 0.5$$
$$H_A : p > 0.5$$

Example 3:

A sociologist wants to test if fewer than 60% of Reno residents are happy. The null and alternative hypotheses are:

$$H_0 : p \geq 0.6$$
$$H_A : p < 0.6$$

3.2 Type I & II Errors

Statistics is not always perfect, and errors are an inevitable part of hypothesis testing. There are two primary types of errors we should be aware of: type I and type II errors. These errors occur when we make incorrect conclusions about the null hypothesis. Let's take a closer look at each type of error:

Type I error	Rejecting a true null hypothesis
Type II error	Failing to reject a false null hypothesis

Type I Error

A **Type I error** occurs when we incorrectly reject the null hypothesis when in reality, it's true.

Let's consider the case of an at-home flu test. The null hypothesis is that you believe you have the flu. If the test result is negative, you reject the idea that you have the flu. If the test was inaccurate, meaning you actually have the flu (and the null hypothesis is true), we have a Type I error on our hands. In this case, the null hypothesis was true but it was mistakenly rejected due to an erroneous at-home test.

Type II Error

A **Type II error** occurs when we fail to reject a null hypothesis that should have been rejected.

Let's consider the case of the at-home flu test again. The null hypothesis is that you believe you have the flu. If the test result is positive, you will not reject the idea that you have the flu. If the test was inaccurate, meaning you actually don't have the flu (and the null hypothesis should have been rejected), we have a Type II error on our hands. In this case, the null hypothesis should have been rejected but it was mistakenly misinterpreted due to an erroneous at-home test.

Practice Examples

Example 1: Court Case

Consider the case where:

H_0: An accused person is innocent. H_A: An accused person is guilty.

What are examples of Type I and Type II errors?

- A **Type I error** occurs when the court wrongly finds the accused person guilty when they are actually innocent. This is rejecting a true null hypothesis.
- A **Type II error** occurs when the court wrongly finds the accused person innocent when they are actually guilty. This is failing to reject a false null hypothesis.

In situations like these, statistical errors can have serious consequences, such as wrongful convictions or acquittals.

Example 2: Disease Screening

Suppose researchers have developed a blood test for a rare disease. The test is fairly reliable: it gives a positive result in 98% of people who have the disease, but it incorrectly gives a positive result in 3% of people who do not have the disease. The null hypothesis is that the individual does not have the disease.

What is the probability of a Type I and Type II error in this scenario?

- A **Type I error** occurs when the test incorrectly indicates that the individual has the disease when, in reality, they do not. This happens in 3% of cases, where the test wrongly gives a positive result for healthy individuals. Thus, the probability of a Type I error is **3%**.
- A **Type II error** occurs when the test incorrectly indicates that the individual does not have the disease when, in reality, they do. This happens in 2% of cases, where the test gives a false negative for those who have the disease. Thus, the probability of a Type II error is **2%**.

By understanding Type I and Type II errors, we can better interpret the results of hypothesis tests and understand the trade-offs involved in making statistical decisions. Type I errors typically lead to overestimation of an effect, while Type II errors lead to underestimation or failure to detect an effect. Both types of errors have consequences, and understanding their likelihood helps in making informed decisions about the reliability of test results.

3.3 Interpreting *P*-Values and Confidence Intervals

When we do an experiment, we want to know if the observed results reflect a true effect or could easily have occurred by chance. the ***p*-value** helps us decide this. The p-value is the probability of observing results as extreme as the ones we've observed in the sample data, assuming the null hypothesis is true.

For example, if the p-value is 0.03, this means that if the null hypothesis were true, we would observe results this extreme only about 3% of the time. Because this probability is low, it provides evidence against the null hypothesis. We can refer to this as **statistical significance**. In hypothesis testing, a small p-value indicates that the data are strong enough to reject the null hypothesis at the chosen significance level (like 0.05).

A **confidence interval** provides a range of values of the true population parameter, like in the case of estimating the difference between two populations. For example, a 95% confidence interval means that if we repeated the study many times and constructed a confidence interval from each sample using the same method, approximately 95% of those intervals would contain the true population value.

In practice, if a 95% confidence interval for the difference between two populations does not include zero [2.5, 7.0], this suggests the difference is statistically significant. We can say we are 95% confident that the true difference lies somewhere between 2.5 and 7.0. This is in the sense that the method used produces intervals that capture the true value 95% of the time in repeated sampling.

4 Hypothesis Testing for the Population Mean and Proportion

4.1 Hypothesis Test for the Population Mean when Population Standard Deviation(σ) Is Known

A **hypothesis test for the population mean** is used when we want to check if the mean of a population is a certain value, given that we only know the sample mean. For example, a coffee shop owner believes that customers spend \$5 or more on coffee each visit. She can implement this hypothesis test to verify the validity of her statement. She collects data from 40 random customers and records how much they spend. After analyzing the data, she finds that the average amount spent is \$4.50 per visit.

She may ask,

"Is this just random chance, or do customers actually spend less than \$5 on average?"

To find out, she performs a hypothesis test!

- If the test shows strong evidence that customers spend less than \$5, she might rethink her pricing or introduce new promotions to encourage higher spending.
- If the test doesn't provide enough evidence, she may stick with her current pricing strategy.

This is how businesses use hypothesis testing, to make data-driven decisions rather than just guessing.

We use this test when the population standard deviation is known. In real data science applications, we are rarely aware of the population standard deviation. However, in the cases that we do, we

We'll use the p-Value approach to conduct this hypothesis test.

The *P*-Value Approach in Six Steps

1. **State the Hypotheses**
   - **Two-tailed test:** $H_0 : \mu = \mu_0$, $H_A : \mu \neq \mu_0$
   - **Right-tailed test:** $H_0 : \mu \leq \mu_0$, $H_A : \mu > \mu_0$
   - **Left-tailed test:** $H_0 : \mu \geq \mu_0$, $H_A : \mu < \mu_0$
2. **Choose a Significance Level** (α)

   Common values: $\alpha = 0.01, 0.05, 0.10$. This represents the permitted probability of a Type I error.
3. **Check Assumptions**

   If population normality is unknown, ensure $n \geq 30$ for the central limit theorem to hold.

   Recall from earlier statistics courses that if a dataset is normally distributed, the mean and median are equal.

4. **Calculate the Test Statistic**

$$z = \frac{\bar{x} - \mu_0}{\sigma/\sqrt{n}}$$

- $\bar{x}$ = mean of the sample data
- μ_0 (pronounced "mew-not") = hypothesized value of the population mean
- σ = standard deviation of the population data
- $\sqrt{n}$ = square root of the number of observations in the sample data

5. **Compute the *p*-value (Using Excel)**

   Remember, a GREATER (>)than sign in the alternative hypothesis means we're doing a RIGHT-tailed test. A LESS (<) than sign in the alternative hypothesis means we're doing a LEFT-tailed test. A $\neq$ sign in the alternative means we are doing a TWO-tailed test.

   - **For a z-test (known variances):**
     - Right-tailed: `=1-NORM.DIST(test statistic, 0, 1, TRUE)`
     - Left-tailed: `=NORM.DIST(test statistic, 0, 1, TRUE)`
     - Two-tailed: `=2*(1-NORM.DIST(ABS(test statistic), 0, 1, TRUE))`

6. **Make a Decision**

   - **Reject** H_0 if p-value $< \alpha$
   - **Fail to reject** H_0 if p-value $\geq \alpha$

Practice Example

Let's do an example. A public health researcher wants to determine if the average age at which individuals receive their first knee replacement surgery is greater than 67, as she currently assumes that the mean age for this procedure is less than 67. Assuming that the data is normally distributed and that the population standard deviation is known to be 9 ($\sigma = 9$), we conduct a hypothesis test. To test this, we take a sample of twenty five patients who have undergone their first knee replacement surgery from a hospital in Reno. We find that the mean age at surgery for this sample is 71.

Step 1, write out the null and alternative hypotheses. Because we assume right now that the mean age is $\leq$ 67, this is our null hypothesis. This is a right-tailed test because we are using the $>$ sign in our hypotheses.

H_0: $\mu \leq 67$.
H_A: $\mu > 67$.

Step 2, choose an α value of 0.01, 0.05, or 0.10. These are commonly referred to as significance levels of 1%, 5%, and 10%, respectively. The α value represents the permitted probability of committing a Type I error. A smaller α value indicates a stricter threshold for statistical significance, reducing the likelihood of false positives but also making it harder to detect true effects.

In practice, researchers typically report results at multiple significance levels (e.g., 1%, 5%, and 10%) to provide a full picture of statistical evidence, allowing readers to assess the strength of the findings. While a 5% significance level ($\alpha = 0.05$) is standard in many fields, more stringent thresholds like 1% are often used in cases where the cost of a false positive is high. For this problem, we'll use $\alpha = 0.01$.

Step 3, verify the population is normally distributed OR that the sample size is $\geq$ 30. We can assume it's normally distributed because it was stated in the question.

Step 4, calculate the test statistic.

$$z = \frac{\bar{x} - \mu_0}{\sigma/\sqrt{n}}$$

Let's remember the definition of each of these terms:

- $\bar{x}$ = mean of the sample data
- μ_0 (pronounced "mew-not") = hypothesized value of the population mean
- σ = population standard deviation
- $\sqrt{n}$ = square root of the number of observations in the sample data

So, let's define each of these terms in this example:

1. $\bar{x} = 71$
2. $\mu_0 = 67$
3. $\sigma = 9$
4. $\sqrt{n} = \sqrt{25}$

After plugging them into the formula:

$$z = \frac{71 - 67}{9/\sqrt{25}} =$$

$$z = \frac{4}{9/5} =$$

$$z = \frac{4}{1.8} =$$
$$z = 2.2222$$

Step 5, using the test statistic of 2.2222, find the *p*-value using Excel. Given the fact that this is a right tailed test because $>$ is used in the alternative hypothesis, we enter $= 1 - NORM.DIST(2.2222, 0, 1, TRUE)$ into Excel and it spits out 0.0131. This is the p-value. This means that if we choose to reject the null hypothesis, there is a 1.31% chance that we are making a Type I error. The p-value is the observed probability of making a type I error. This seems like a small chance, but is it small enough to allow us to reject the null hypothesis? This is where the α values come into play. The α value is the permitted probability of making a Type I error.

Step 6, using the *p*-value and α value, either reject or do not reject the null hypothesis. Reject the null if p-value < α. Do not reject the null if p-value $\geq \alpha$. In this case, our p-value is 0.0131 and our chosen α is 0.01. Therefore, 0.0131 > 0.01 so we do not reject the null hypothesis at the 1% significance level.

If we do not reject the null hypothesis, this suggests that we do not have sufficient statistical evidence to conclude that the mean age at which individuals undergo their first knee replacement surgery is greater than 67. This does not explicitly translate to the true mean age is 67 or lower; rather, our sample data does not provide strong enough evidence to support a conclusion that the mean age is higher at the chosen significance level.

In the context of healthcare, this result implies that while the mean age for the sample was 71, we cannot confidently claim that patients are waiting longer on average for their first knee replacement surgery compared to what was hypothesized (age 67). This could indicate that, while there may be variability in the timing of surgeries across different individuals, factors such as access to healthcare, insurance coverage, or other health conditions may be influencing the timing of surgeries rather than simply a trend toward later surgeries.

Given this outcome, further investigation may be needed, perhaps with a larger sample size or by examining other demographic and clinical variables, to better understand the factors that contribute to the age at which knee replacement surgeries are performed.

4.2 Hypothesis Test for the Population Mean When Standard Deviation(σ) Is Unknown

We've been talking about how to conduct a hypothesis test for the population mean (μ) when we **know** the population standard deviation (σ). But in the real world, we usually don't know σ. In this unit, we'll learn how to conduct a hypothesis test for the population mean when we don't know the population standard deviation. In any given word problem, if you are only provided the sample standard deviation and not the population standard deviation, that means σ is unknown and therefore you should follow these steps.

The *P*-Value Approach in 6 Steps

1. **State the Hypotheses**

   - **Two-tailed test:** $H_0 : \mu = \mu_0$, $H_A : \mu \neq \mu_0$
   - **Right-tailed test:** $H_0 : \mu \leq \mu_0$, $H_A : \mu > \mu_0$
   - **Left-tailed test:** $H_0 : \mu \geq \mu_0$, $H_A : \mu < \mu_0$

2. **Choose a Significance Level** (α)

   Common values: $\alpha = 0.01, 0.05, 0.10$. This represents the permitted probability of a Type I error.

3. **Check Assumptions**

   If population normality is unknown, ensure $n \geq 30$ for the central limit theorem to hold.

   Recall from earlier statistics courses that if a dataset is normally distributed, the mean and median are equal.

4. **Calculate test statistic (also known as a t value).**

$$t_{df} = \frac{\bar{x} - \mu_0}{s/\sqrt{n}}$$

Definitions:

- $\bar{x}$ = mean value of our sample data
- μ_0 (pronounced "mew-not") = hypothesized value of the population mean
- s = standard deviation of the sample
- n = number of observations in the sample
- $df = n - 1$ = number of observations in the sample - 1. This is also known as the "degrees of freedom" which I will talk more about in below sections.

5. **Compute the *p*-value (Using Excel)**

   Remember, a GREATER (>)than sign in the alternative hypothesis means we're doing a RIGHT-tailed test. A LESS (<) than sign in the alternative hypothesis means we're doing a LEFT-tailed test. A $\neq$ sign in the alternative means we are doing a TWO-tailed test.

   - **For a z-test (known variances):**
     - Right-tailed: `=T.DIST.RT(test statistic, df)`
     - Left-tailed: `=1-T.DIST.RT(test statistic, df)`
     - Two-tailed: `=2*T.DIST.RT(ABS(test statistic), df)`

6. **Make a Decision**

   - **Reject** H_0 if p-value $< \alpha$
   - **Fail to reject** H_0 if p-value $\geq \alpha$

Let's do an example. A university dean thinks that students study $\geq$ 24 hours/week. She is curious if they study less than that. She randomly selects a sample of thirty five students and interviews them on their quantity of study hours per week. From these interviews, she calculates a sample mean of 16.3714 hours and a sample standard deviation of 7.2155. **Remember, this is only the standard deviation of the sample, not the entire university population.** In statistics, s is used to denote the standard deviation of the sample and σ denotes the standard deviation of the population data.

Step 1, write out the null and alternative hypotheses. Because we assume right now that the mean study time $\geq$ 24 hours/week, this is our null hypothesis. This is a left-tailed test because we are using the < sign in our alternative hypothesis.

$$H_0\text{: } \mu \geq 24.$$
$$H_A\text{: } \mu < 24.$$

Step 2, choose an α value of 0.01, 0.05, or 0.10. These are commonly referred to as significance levels of 1%, 5%, and 10%, respectively. The α value represents the permitted probability of committing a Type I error. A smaller α value indicates a stricter threshold for statistical significance, reducing the likelihood of false positives but also making it harder to detect true effects. For this example, let's choose an $\alpha = 0.05$.

Step 3, verify that the population is normally distributed OR verify that the sample size is sufficiently large, which is $\geq$ 30. In this case, our sample size is 35, and 35 > 30.

Step 4, calculate the test statistic.

$$t_{df} = \frac{\bar{x} - \mu_0}{s/\sqrt{n}}$$

Let's recall what each of these terms mean:

- $\bar{x}$ = mean value of our sample data
- μ_0 (pronounced "mew-not") = hypothesized value of the population mean
- s = standard deviation of the sample
- n = number of observations in the sample
- $df = n - 1$ = number of observations in the sample - 1. This is also known as the "degrees of freedom". The degrees of freedom are a fancy way of measuring how many pieces went into computing the sample mean. It is important to note that the degrees of freedom vary depending on the test. For example, some df are computed as $n - 2$.

So, let's define each of these terms in this example:

- $\bar{x} = 16.3714$
- $\mu_0 = 24$
- $s = 7.2155$
- $\sqrt{n} = \sqrt{35}$
- $n - 1 = 35 - 1 = 34$

After plugging them into the formula:

$$t_{df} = \frac{16.3714 - 24}{7.2155/\sqrt{35}}$$

$$t_{df} = \frac{-7.6286}{7.2155/5.9161}$$

$$t_{df} = \frac{-7.6286}{1.2196}$$

$$t_{df} = -6.2550$$

Step 5, using the test statistic of -6.2550, find the *p*-value using Excel. We enter $= 1 - T.DIST.RT(-6.2550, 34)$ into Excel and it yields 2.013E-07. This translates to .0000002013, a very tiny number. This is the *p*-value. This means that if we choose to reject the null hypothesis, there is only a 0.00002013% chance that we are wrong.

Step 6, using the *p*-value and α value, either reject or do not reject the null hypothesis. Reject the null if *p*-value < α. Do not reject the null if *p*-value $\geq \alpha$. In this case, our *p*-value is

.0000002013 and our chosen α is .05. Therefore, .0000002013 < .05 and we reject the null hypothesis at the 5% significance level.

If we reject the null hypothesis, this suggests that there is sufficient statistical evidence to conclude that the average number of study hours per week for the population of students at the university is NOT greater than 24 hours. We can favor the conclusion that students are studying less than 24 hours per week.

From a practical standpoint, this outcome may indicate that the assumption of 24 hours is too high for the general student population. These findings could prompt the dean and the university to explore ways to encourage more study time, such as improving academic support services.

4.3 Hypothesis Test for the Population Proportion

A hypothesis test for a population proportion is used when we want to determine if the proportion of a certain characteristic in a population differs from a claimed or expected value. Instead of testing a mean, we are testing a percentage or fraction of individuals in a group who have a specific trait. This type of test is useful in many real-world applications, such as evaluating customer satisfaction rates, voting preferences, or product defect rates.

For example, a toy store manager believes that at least 70% of customers are happy with a new stuffed animal. However, she wants to make sure before ordering more!

She surveys one hundred random customers who purchased the toy and finds that only **65%** say they are happy with it.

Now, she asks:

> "Is this just random chance, or are fewer than 70% of customers actually happy?"

To find out, she performs a hypothesis test for the proportion!

- If the test shows strong evidence that fewer than 70% of customers are satisfied, she might improve the toy's design or change her marketing strategy.
- If the test doesn't provide enough evidence, she continues selling the toy as planned.

The *P*-Value Approach in Six Steps

1. **State the Hypotheses**
   - **Two-tailed test:** $H_0 : p = p_0$, $H_A : p \neq p_0$
   - **Right-tailed test:** $H_0 : p \leq p_0$, $H_A : p > p_0$
   - **Left-tailed test:** $H_0 : p \geq p_0$, $H_A : p < p_0$
2. **Choose a Significance Level** (α)

   Common values: $\alpha = 0.01, 0.05, 0.10$. This represents the permitted probability of a Type I error.
3. **Check Assumptions**

   Verify that the population data is normally distributed OR that the sample size is large enough.

   To verify the sample size is large enough, we need to check that $n * p_0 \geq 5$, where n = sample size and p_0 = hypothesized proportion value.
4. **Calculate test statistic (also known as a z value).**

$$z = \frac{\bar{p} - p_0}{\sqrt{p_0(1 - p_0)/n}}$$

Let's break down the definition of each of these terms:

- $\bar{p}$ (pronounced "p-bar") = proportion of the sample
- p_0 (pronounced "p-not") = hypothesized value of the population proportion
- n = number of observations in the sample

5. **Compute the *p*-value (Using Excel)**

   Remember, a GREATER (>)than sign in the alternative hypothesis means we're doing a RIGHT-tailed test. A LESS (<) than sign in the alternative hypothesis means we're doing a LEFT-tailed test. A $\neq$ sign in the alternative means we are doing a TWO-tailed test.

   - **For a z-test (known variances):**
     - Right-tailed: `=1-NORM.DIST(test statistic, 0, 1, TRUE)`
     - Left-tailed: `=NORM.DIST(test statistic, 0, 1, TRUE)`
     - Two-tailed: `=2*(1-NORM.DIST(ABS(test statistic), 0, 1, TRUE))`

6. **Make a Decision**

   - **Reject** H_0 if p-value $< \alpha$
   - **Fail to reject** H_0 if p-value $\geq \alpha$

Let's do an example. A grocery store chain currently believes that more than 40% of its customers have changed their shopping habits due to rising food prices. A recent survey of 180 customers finds that sixty seven customers have adjusted their shopping habits to cope with increasing costs.

Step 1, write out the null and alternative hypotheses. Because we assume right now that the status quo for the population is that more than 40% of customers have changed their shopping habits due to rising food prices, this is the null hypothesis. This is a left-tailed test because we are using the $\leq$ sign in our alternative hypothesis.

$$H_0: p > 0.40$$
$$H_A: p \leq 0.40.$$

Step 2, choose an α value of 0.01, 0.05, or 0.10. These are commonly referred to as "significance levels" of 1%, 5%, and 10%, respectively. The α value represents the permitted probability of rejecting the null hypothesis when it is actually true (Type I error). A smaller α value indicates a stricter threshold for statistical significance, reducing the likelihood of false positives but also making it harder to detect true effects. For this example, let's choose an $\alpha = 0.05$.

Step 3, verify that the population data is normally distributed OR that the sample size is large enough. We have not been given the population data in this case. We only have access to the sample data. Therefore, we need to verify that the sample size is large enough. To do this, we just need to check that $np_0 \geq 5$, where n = sample size and p_0 = hypothesized proportion value. In this case, n = 180 and p_0 = 40% or 0.4. So, np_0 = 180 $*$ .40 = 72 $\geq$ 5.

Step 4, calculate the test statistic.

$$z = \frac{\bar{p} - p_0}{\sqrt{p_0(1 - p_0)/n}}$$

Let me break down what all these terms mean:

- $\bar{p}$ (pronounced "p-bar") is the proportion of our sample (67/180)
- p_0 (pronounced "p-not") is our hypothesized value of the population average (we think more than 40% of households make lifestyle changes due to environmental concerns.)
- n is the number of observations in our sample (they interviewed 180 people).

So, let's define each of these terms in this example:

- $\bar{p}$ = 67/180 = 0.3722
- p_0 = 0.4
- n = 180

We plug them into the equation to look like this:

$$z = \frac{0.3722 - 0.4}{\sqrt{0.4(1 - 0.4)/180}}$$

$$z = \frac{-0.0278}{\sqrt{0.4(0.6)/180}}$$

$$z = \frac{-0.0278}{\sqrt{0.24/180}}$$

$$z = \frac{-0.0278}{\sqrt{0.0013}}$$

$$z = \frac{-0.0278}{0.0365}$$

$$z = \frac{-0.0278}{0.0361}$$

$$z = -0.7613$$

Step 5, using the test statistic of -0.7613, find the p-value using Excel. We enter $= NORM.DIST(-0.7613, 0, 1, TRUE)$ into Excel and it produces 0.2232. This means that if we choose to reject the null hypothesis, there is a 22.32% chance that we are wrong.

Step six, using the p and α values, either reject or do not reject the null hypothesis. Reject the null if p-value < α. Do not reject the null if p-value $\geq \alpha$. In this case, our p-value is 0.2232 and our chosen α is .05. Therefore, 0.2232 > .05 and we do NOT reject the null hypothesis at the 5% significance level.

At the 5% significance level, we **do not have enough statistical evidence** to conclude that the proportion of households that have changed their shopping habits due to rising food prices is **less than 40%**. This means that the data does not provide strong enough support to challenge the claim that at least 40% of households have made changes. The observed difference between the sample proportion and 40% could be due to **random sampling variability**, meaning that the sample we selected may not perfectly represent the entire population. In other words, the sample might have happened to include a higher or lower proportion of households that made lifestyle changes, even though the true population proportion is closer to or exactly 40%.

However, failing to reject H_0 does **not** prove that the true proportion is greater than or equal to 40%—it simply means we don't have enough evidence to conclude otherwise. Future research with a larger or more diverse sample could help clarify whether the observed difference is truly reflective of the population or just a result of sampling variation.

5 Hypothesis Testing for Differences between Two Means

5.1 Control and Treatment Groups (Independent t-test)

A hypothesis test for the difference between two means is used to determine whether there is a statistically significant difference between the average values of two groups. This test is commonly applied in experimental and observational studies where researchers compare outcomes between different populations.

There are two primary study designs: matched sampling and control/treatment group comparisons. In **matched sampling** (paired t-test), each subject serves as their own control, such as in pre-test and post-test studies, which reduces variability and increases statistical power. In contrast, a **control/treatment group** scenario (independent t-test) compares two distinct groups, such as patients receiving a new medication versus those receiving a placebo. For example, in a clinical trial evaluating a new cholesterol-lowering drug, researchers might compare the mean LDL cholesterol levels between a treatment group receiving the drug and a control group receiving a placebo to determine if the medication has a significant effect.

The *P*-Value Approach in Six steps:

1. **State the Hypotheses**
   - **Two-tailed test:** $H_0 : \mu_1 - \mu_2 = 0$, $H_A : \mu_1 - \mu_2 \neq 0$
   - **Right-tailed test:** $H_0 : \mu_1 - \mu_2 \leq 0$, $H_A : \mu_1 - \mu_2 > 0$
   - **Left-tailed test:** $H_0 : \mu_1 - \mu_2 \geq 0$, $H_A : \mu_1 - \mu_2 < 0$
2. **Choose a Significance Level** (α) Common values: $\alpha = 0.01, 0.05, 0.10$. This represents the permitted probability of a Type I error.
3. **Check Assumptions** If population normality is unknown, ensure $n_1 \geq 30$ and $n_2 \geq 30$ for the central limit theorem to hold.
4. **Calculate the Test Statistic**
   - **If population variances are known:**

$$z = \frac{\bar{x_1} - \bar{x_2}}{\sqrt{\frac{\sigma_1^2}{n_1} + \frac{\sigma_2^2}{n_2}}}$$

   - **If population variances are unknown but assumed equal:**

$$t_{df} = \frac{\bar{x_1} - \bar{x_2}}{S_p^2\sqrt{\frac{1}{n_1} + \frac{1}{n_2}}},$$

where

$$S_p^2 = \sqrt{\frac{(n_1 - 1)s_1^2 + (n_2 - 1)s_2^2}{n_1 + n_2 - 2}}.$$

where degrees of freedom are approximated as:

$$df = n_1 + n_2 - 2$$

- **If population variances are unknown and assumed unequal:**

$$t_{df} = \frac{\bar{x}_1 - \bar{x}_2}{\sqrt{\frac{s_1^2}{n_1} + \frac{s_2^2}{n_2}}},$$

with degrees of freedom approximated as:

$$df = \frac{(s_1^2/n_1 + s_2^2/n_2)^2}{(s_1^2/n_1)^2/(n_1 - 1) + (s_2^2/n_2)^2/(n_2 - 1)}.$$

Definitions:

- $\bar{x}_1, \bar{x}_2$ = Sample means for groups 1 and 2.
- σ_1^2, σ_2^2 = Population variances for groups 1 and 2.
- s_1^2, s_2^2 = Sample variances for groups 1 and 2.
- n_1, n_2 = Sample sizes for groups 1 and 2.
- S_p^2 = Pooled sample variance.
- t_{df} = Test statistic under the t-distribution with degrees of freedom df.

5. **Compute the *p*-value (Using Excel)**

   - **For a z-test (known variances):**
     - Right-tailed: `=1-NORM.DIST(test statistic, 0, 1, TRUE)`
     - Left-tailed: `=NORM.DIST(test statistic, 0, 1, TRUE)`
     - Two-tailed: `=2*(1-NORM.DIST(ABS(test statistic), 0, 1, TRUE))`
   - **For a *t*-test (unknown variances):**
     - Right-tailed: `=T.DIST.RT(test statistic, df)`
     - Left-tailed: `=1-T.DIST.RT(test statistic, df)`
     - Two-tailed: `=2*T.DIST.RT(ABS(test statistic), df)`

6. **Make a Decision**

   - **Reject** H_0 if p-value $< \alpha$
   - **Fail to reject** H_0 if p-value $\geq \alpha$

Let's do an example. An economist currently believes the average weekly food expenses for households in City 1 is less than or equal to the average weekly food expenses for households in City 2. She surveys thirty five households in City 1 that yield an average weekly food expense of \$164. She surveys thirty households in City 2 that yield an average weekly food expense of \$159. The standard deviation for the population of City 1 households is \$12.5 and for City 2 is \$9.25. At the 5% significance level, can we support the economist's claim that City 1 has reduced expenses than City 2?

1. **State the Hypotheses**

    - Given that the economist believes City 1 has lower expenses than City 2, we set up the hypotheses as follows:

$$H_0 : \mu_1 - \mu_2 \leq 0$$
$$H_A : \mu_1 - \mu_2 > 0$$

2. **Choose a Significance Level** (α)

    - We are given $\alpha = 0.05$.

3. **Check Assumptions**

    - Since the population normality is unknown, we check if $n_1 \geq 30$ and $n_2 \geq 30$. Here, $n_1 = 35$ and $n_2 = 30$, so the central limit theorem applies, and we can proceed with a normal approximation.

4. **Calculate the Test Statistic**

    Since population variances are known, we use the z-test formula:

$$z = \frac{\bar{x_1} - \bar{x_2} - d_0}{\sqrt{\frac{\sigma_1^2}{n_1} + \frac{\sigma_2^2}{n_2}}}$$

Definitions:

- $\bar{x_1}$ = Sample mean for City 1 = 164
- $\bar{x_2}$ = Sample mean for City 2 = 159
- d_0 = Hypothesized mean difference = 0
- σ_1^2 = Population variance for City 1 = 12.50^2
- σ_2^2 = Population variance for City 2 = 9.25^2
- n_1 = Sample size for City 1 = 35
- n_2 = Sample size for City 2 = 30

Plugging in the values:

$$z = \frac{164 - 159 - 0}{\sqrt{\frac{12.50^2}{35} + \frac{9.25^2}{30}}}$$

Computing:

$$z = \frac{5}{\sqrt{\frac{156.25}{35} + \frac{85.5625}{30}}}$$

$$z = \frac{5}{\sqrt{4.4643 + 2.8521}}$$

$$z = \frac{5}{\sqrt{7.3164}}$$
$$z = \frac{5}{2.705}$$
$$z \approx 1.85$$

5. **Compute the *p*-value (Using Excel)**
   - Since this is a right-tailed test, we use:

```
=1-NORM.DIST(1.85, 0, 1, TRUE)
```

   - This yields $p \approx 0.032$.

6. **Make a Decision**
   - Since $p = 0.032 < \alpha = 0.05$, we reject H_0.
   - Conclusion: There is sufficient evidence at the 5% significance level to reject the hypothesis that City 1 has lower average weekly food expenses than City 2.

5.2 Matched Sampling (Paired t-test)

A hypothesis test for the difference between two means in a paired design is used to determine whether there is a statistically significant difference between the means of two related groups. This test is often applied in experiments where subjects are measured twice, such as in pre-test and post-test studies.

In the case of **matched sampling** (paired t-test), each subject serves as their own control. This reduces variability and increases statistical power by comparing the differences within the same individuals over time or between conditions. For example, in a clinical trial evaluating the effect of a new drug, researchers might compare blood pressure levels in patients before and after receiving the treatment to determine if there is a significant reduction in blood pressure.

The *P*-Value Approach in Six steps:

1. **State the Hypotheses**
   - **Two-tailed test:** $H_0 : \mu_D = 0$, $H_A : \mu_D \neq 0$
   - **Right-tailed test:** $H_0 : \mu_D \leq 0$, $H_A : \mu_D > 0$
   - **Left-tailed test:** $H_0 : \mu_D \geq 0$, $H_A : \mu_D < 0$
2. **Choose a Significance Level** (α) Common values: $\alpha = 0.01, 0.05, 0.10$. This represents the permitted probability of a Type I error.
3. **Check Assumptions**
   - Ensure that the differences between paired observations are approximately normally distributed. If the sample size is large ($n \geq 30$), the central limit theorem allows for the use of a normal approximation.
4. **Calculate the Test Statistic**
   - The test statistic for a paired t-test is given by:

   $$t = \frac{\bar{D}}{S_D/\sqrt{n}},$$

   where:
   - $\bar{D}$ = Mean of the differences between paired observations,
   - S_D = Standard deviation of the differences between paired observations
   - n = Number of pairs (sample size).
5. **Compute the *p*-value (Using Excel)**
   - For a right-tailed test:

   ```
   =T.DIST.RT(test statistic, n-1)
   ```

   - For a left-tailed test:

```
=1-T.DIST.RT(test statistic, n-1)
```

- For a two-tailed test:

```
=2*T.DIST.RT(ABS(test statistic), n-1)
```

6. **Make a Decision**
   - **Reject** H_0 if p-value $< \alpha$
   - **Fail to reject** H_0 if p-value $\geq \alpha$

Let's do an example. A researcher wants to evaluate the effect of a new exercise program on the weight loss of forty participants. Each participant's weight is measured before and after completing the program. The researcher is skeptical and currently believes that the difference before and after participation in the program is negative (Pre Weight - Post Weight < 0). The mean of the differences in weight before and after the program for all 40 participants is 2.85. The standard deviation of these weight differences is 0.62. We will perform a paired t-test at the 5% level to determine if there is a statistically significant difference in weight loss after the exercise program.

1. **State the Hypotheses**
   - Given that the researcher believes the exercise program results in weight gain (a negative difference), we set up the hypotheses as follows:

$$H_0 : \mu_D \leq 0$$
$$H_A : \mu_D > 0$$

2. **Choose a Significance Level** (α)
   - We are given that $\alpha = 0.05$.
3. **Check Assumptions**
   - The sample size is large ($n = 40$), so by the Central Limit Theorem, the sampling distribution of the differences is approximately normal.
4. **Calculate the Test Statistic**

$$t = \frac{2.85}{0.62/\sqrt{40}} \approx \frac{2.85}{0.098} \approx 29.08$$

5. **Compute the p-value (Using Excel)**
   - Since this is a right-tailed test, we use:

```
=T.DIST.RT(29.08, 39)
```

   - This yields $p \approx 0.000$.

6. **Make a Decision**

- Since $p = 0.000 < \alpha = 0.05$, we reject H_0.
- Conclusion: There is sufficient evidence at the 5% significance level to support the hypothesis that the exercise program does NOT result in weight gain.

6 Hypothesis Testing for Difference between Two Proportions

6.1 Inference of the Difference between Two Proportions

A hypothesis test for the difference between two proportions is used to determine whether there is a statistically significant difference between the proportions of two independent groups. This test is commonly applied when comparing the success rates of two different treatments or interventions, such as comparing the effectiveness of two drugs.

In the case of **independent samples** (difference of proportions test), we compare two proportions derived from two independent groups. This test is often used in studies where the data consists of success/failure outcomes, such as the proportion of people who recover from an illness under two different treatment conditions.

We'll learn the *P*-Value and confidence interval methods for this hypothesis test.

The *P*-Value Approach in Six Steps:

1. **State the Hypotheses**

   - **Two-tailed test:**

   $$H_0: p_1 - p_2 = 0, \quad H_A: p_1 - p_2 \neq 0$$

   - **Right-tailed test:**

   $$H_0: p_1 - p_2 \leq 0, \quad H_A: p_1 - p_2 > 0$$

   - **Left-tailed test:**

   $$H_0: p_1 - p_2 \geq 0, \quad H_A: p_1 - p_2 < 0$$

2. **Choose a Significance Level** (α) Common values: $\alpha = 0.01, 0.05, 0.10$. This represents the permitted probability of a Type I error.

3. **Check Assumptions**

   - Both sample sizes should be large enough to approximate the sampling distribution of the difference in proportions by a normal distribution. A common rule of thumb is to ensure all of the following statements are true, where p_1 and p_2 are the sample proportions.

   $$\begin{gathered} n_1 p_1 \geq 10 \\ n_1(1 - p_1) \geq 10 \\ n_2 p_2 \geq 10 \\ n_2(1 - p_2) \geq 10 \end{gathered}$$

4. **Calculate the Test Statistic**

   - The test statistic for the difference between two proportions is given by:

   $$z = \frac{(\hat{p}_1 - \hat{p}_2)}{\sqrt{\hat{p}(1 - \hat{p})\left(\frac{1}{n_1} + \frac{1}{n_2}\right)}},$$

   where:

- $\hat{p}_1$ = Sample proportion from group 1,
- $\hat{p}_2$ = Sample proportion from group 2,
- $\hat{p}$ = Pooled sample proportion: $\hat{p} = \frac{x_1+x_2}{n_1+n_2}$, where x_1 and x_2 are the number of "successes" in each group,
- n_1 and n_2 = Sample sizes of groups 1 and 2.

5. **Compute the *p*-value (Using Excel)**
   - Right-tailed: `=1-NORM.DIST(test statistic, 0, 1, TRUE)`
   - Left-tailed: `=NORM.DIST(test statistic, 0, 1, TRUE)`
   - Two-tailed: `=2*(1-NORM.DIST(ABS(test statistic), 0, 1, TRUE))`
6. **Make a Decision**
   - **Reject** H_0 if p-value $< \alpha$
   - **Fail to reject** H_0 if p-value $\geq \alpha$

The Confidence Interval Approach in Six Steps:

1. **State the Hypotheses**

$$H_0 : p_1 - p_2 = 0, \quad H_A : p_1 - p_2 \neq 0$$

2. **Choose a Confidence Level**

   Common values include:

$$\text{Confidence Level} = 90\%, 95\%, 99\%$$

   These correspond to significance levels $\alpha = 0.10, 0.05, 0.01$, respectively, where $\alpha = 1 -$ Confidence Level.

3. **Check Assumptions**
   - Both sample sizes should be large enough to approximate the sampling distribution of the difference in proportions by a normal distribution. A common rule of thumb is for the following statements to be true: $n_1p_1 \geq 10$, $n_1(1-p_1) \geq 10$, $n_2p_2 \geq 10$, and $n_2(1-p_2) \geq 10$, where p_1 and p_2 are the sample proportions.
4. **Calculate the Confidence Interval**
   - The confidence interval for the difference between two proportions is given by:

$$(\hat{p}_1 - \hat{p}_2) \pm z_{\alpha/2} \times \sqrt{\hat{p}(1-\hat{p})\left(\frac{1}{n_1} + \frac{1}{n_2}\right)},$$

   where:

- $\hat{p}_1$ = Sample proportion from group 1,
- $\hat{p}_2$ = Sample proportion from group 2,
- $\hat{p}$ = Pooled sample proportion: $\hat{p} = \frac{x_1+x_2}{n_1+n_2}$, where x_1 and x_2 are the number of successes in each group,
- n_1 and n_2 = Sample sizes of groups 1 and 2,
- $z_{\alpha/2}$ = Z-score corresponding to the desired confidence level (e.g., for 95%, $z_{\alpha/2} = 1.96$).

5. **Interpret the Confidence Interval**
   - If the interval contains 0, fail to reject H_0, as there is no significant difference between the two proportions.
   - If the interval does not contain 0, reject H_0, indicating a significant difference between the two proportions.

6. **Make a Decision**
   - **Reject** H_0 if the confidence interval does not contain 0.
   - **Fail to reject** H_0 if the confidence interval contains 0.

Example Problem:
Suppose we are interested in comparing the proportions of individuals in two cities who are satisfied with their public transportation systems. Conduct a two-tailed hypothesis test at $\alpha = 0.05$ using both the p-value and confidence interval approaches.

Details of the Sample:

- **City A:** A random sample of 300 individuals was selected from City A. Out of these, 216 individuals expressed satisfaction with the public transportation system. This means that the proportion of satisfied individuals in City A is:

$$\hat{p}_1 = \frac{216}{300} = 0.72$$

- **City B:** A random sample of 250 individuals was selected from City B. Out of these, 170 individuals expressed satisfaction with the public transportation system. This means that the proportion of satisfied individuals in City B is:

$$\hat{p}_2 = \frac{170}{250} = 0.68$$

***P*-Value Approach**

Step 1: State the Hypotheses

- We are performing a two-tailed test.

$$H_0 : p_1 - p_2 = 0, \quad H_A : p_1 - p_2 \neq 0$$

where: p_1 is the proportion of satisfied individuals in City 1, and p_2 is the proportion of satisfied individuals in City 2.

Step 2: Choose a Significance Level

Let the significance level be $\alpha = 0.05$.

Step 3: Check Assumptions

- The sample sizes are sufficiently large because all of the following statements are true:
- $n_1 = 300$
- $n_2 = 250$
- $n_1 * p_1 = 300 * 0.72 = 216 > 10$
- $n_1 * (1 - p_1) = 300 * (1 - 0.72) = 84 > 10$
- $n_2 * p_2 = 250 * 0.68 = 170 > 10$
- $n_2 * (1 - p_2) = 250 * (1 - 0.68) = 80 > 10$

Step 4: Calculate the Test Statistic

- The sample proportions are $\hat{p_1} = 0.72$ and $\hat{p_2} = 0.68$.
- The pooled sample proportion is:

$$\hat{p} = \frac{x_1 + x_2}{n_1 + n_2} = \frac{216 + 170}{300 + 250} = 0.70$$

- The test statistic is calculated as:

$$z = \frac{\hat{p_1} - \hat{p_2}}{\sqrt{\hat{p}(1 - \hat{p}) \left(\frac{1}{n_1} + \frac{1}{n_2}\right)}} = \frac{0.72 - 0.68}{\sqrt{0.70 \times 0.30 \left(\frac{1}{300} + \frac{1}{250}\right)}} \approx 1.14$$

Step 5: Compute the *p*-value

- Since this is a two-tailed test, we use:

```
=2*(1-NORM.DIST(ABS(1.14),0,1,TRUE))
```

- This yields $p \approx 0.254$.

Step 6: Make a Decision

- Since the p-value (0.254) is greater than the significance level $\alpha = 0.05$, we fail to reject the null hypothesis.

Confidence Interval Approach

Step 1: State the Hypotheses

We are performing a two-tailed test.

$$H_0 : p_1 - p_2 = 0, \quad H_A : p_1 - p_2 \neq 0$$

Step 2: Choose a Confidence Level

Let the confidence level be 95%, which corresponds to a significance level of $\alpha = 0.05$.

Step 3: Check Assumptions

The sample sizes are sufficiently large, and the conditions for normality are met as discussed above.

Step 4: Calculate the Confidence Interval

The confidence interval for the difference between proportions is calculated as:

$$(\hat{p_1} - \hat{p_2}) \pm z_{\alpha/2} \times \sqrt{\hat{p}(1-\hat{p})\left(\frac{1}{n_1} + \frac{1}{n_2}\right)}$$

where $z_{\alpha/2}$ for a 95% confidence level is 1.96.

Thus, the confidence interval is:

$$(0.72 - 0.68) \pm 1.96 \times \sqrt{0.70 \times 0.30\left(\frac{1}{300} + \frac{1}{250}\right)} = 0.04 \pm 0.057$$

The confidence interval is $(-0.017, 0.097)$.

Step 5: Interpret the Confidence Interval

Since the confidence interval includes 0, we fail to reject the null hypothesis.

Step 6: Make a Decision

Because the confidence interval contains 0, we fail to reject the null hypothesis, which indicates no significant difference between the two proportions.

7 Analysis of Variance (ANOVA) Testing

7.1 One-Way ANOVA Test

One-Way ANOVA testing evaluates whether the means of three or more independent populations demonstrate a statistically significant difference. For example, suppose we assign volunteers into three groups (A, B, and C), each using a different running shoe, and measure their mile times. We then compare the group means to determine if observed differences are statistically significant or due to chance. Although it would be possible to perform multiple pairwise tests (A vs. B, B vs. C, etc), this increases the risk of Type I errors. One-Way ANOVA mitigates this by testing all groups simultaneously.

Assumptions

1. Populations are normally distributed.
2. Population standard deviations are unknown but assumed equal.
3. Samples are independently selected.

Steps for Conducting a One-Way ANOVA Test

1. **State hypotheses:**
   - Null hypothesis: $H_0 : \mu_1 = \mu_2 = \cdots = \mu_c$
   - Alternative hypothesis: At least one mean differs.
2. **Choose** α (e.g., 0.01, 0.05, or 0.10). This is the permitted probability of a Type I error.
3. **Calculate the grand mean:**

   $$\bar{\bar{x}} = \frac{1}{n_T} \sum_{i=1}^{c} \sum_{j=1}^{n_i} x_{ij}$$

   where:

   - c is the number of groups,
   - n_i is the sample size for group i,
   - n_T is the total number of observations across all groups,
   - x_{ij} represents each individual observation.
4. **Compute the Sum of Squares Between Groups (SSTR), which represents the total variation due to the differences between group means:**

   $$SSTR = \sum_{i=1}^{c} n_i (\bar{x}_i - \bar{\bar{x}})^2$$

   where:

- $\bar{x}_i$ is the sample mean for group i,
- n_i is the sample size for group i,
- $\bar{\bar{x}}$ is the grand mean.

5. **Compute Mean Square Between Groups (MSTR), the average variation between the group means:**

$$MSTR = \frac{SSTR}{c-1}$$

where:

- c is the number of groups.

6. **Compute Sum of Squares Within Groups (SSE), the variation of individual observations around the group mean:**

$$SSE = \sum_{i=1}^{c}(n_i - 1)s_i^2$$

where:

- s_i^2 is the sample variance for group i,
- n_i is the sample size for group i.

7. **Compute Mean Square Error (MSE), the average within-group variation:**

$$MSE = \frac{SSE}{n_T - c}$$

where:

- c is the number of groups,
- n_T is the total number of observations across all groups.

8. **Compute the test statistic and degrees of freedom:**

$$F = \frac{MSTR}{MSE}$$

$$df_1 = c - 1$$

$$df_2 = n_T - c$$

where:

- c is the number of groups,
- n_T is the total number of observations across all groups.

9. **Compute the p-value:**

- In Excel, use the command: `=F.DIST.RT(F, df1, df2).`

10. **Decision rule and interpretation:**

    - If p-value $< \alpha$, reject H_0 (significant difference exists among group means).
    - If p-value $\geq \alpha$, do not reject H_0 (no significant difference).
    - Interpretation: If we reject H_0, we conclude that at least one treatment group differs significantly. If we fail to reject H_0, there is insufficient evidence to claim a difference in treatment effects.

The output of an ANOVA test is traditionally presented like in the table structure below:

ANOVA Table Structure

Source of Variation	Sum of Squares (SS)	Degrees of Freedom (df)	Mean Square (MS)	F-Statistic
Between Groups	SSTR	$c-1$	MSTR	$F = \frac{MSTR}{MSE}$
Within Groups	SSE	$n_T - c$	MSE	
Total	$SSTR + SSE$	$n_T - 1$		

Table 7.1: ANOVA table

Example: Patient Recovery Times

Suppose we analyze the recovery times of patients undergoing three different physical therapy treatments at a 5% significance level. There are three groups of ten patients each, totaling thirty observations.

Treatment Group	Sample Mean (Days)	Sample Variance	Sample Size
Treatment A	10.2	4.5	10
Treatment B	12.5	5.1	10
Treatment C	9.8	3.8	10

Table 7.2: Patient recovery times by treatment group

The hypotheses are:

$$H_0\colon \mu_1 = \mu_2 = \mu_3$$
$$H_a\colon \text{At least one mean differs.}$$

Given the sample means $\bar{x}_1 = 10.2$ days, $\bar{x}_2 = 12.5$ days, and $\bar{x}_3 = 9.8$ days, the grand mean is:

$$\bar{\bar{x}} = \frac{10.2 + 12.5 + 9.8}{3} = 10.83$$

Now, we compute:

$$SSTR = 10(10.2 - 10.83)^2 + 10(12.5 - 10.83)^2 + 10(9.8 - 10.83)^2 = 42.47$$

$$MSTR = \frac{42.47}{3-1} = 21.23$$

$$SSE = (10-1)(4.5) + (10-1)(5.1) + (10-1)(3.8) = 120.6$$

$$MSE = \frac{120.6}{30-3} = 4.47$$

$$F = \frac{21.23}{4.47} = 4.75$$

Using Excel: `=F.DIST.RT(4.75, 2, 27)`, the p-value is approximately 0.017.

Since $p < 0.05$, we reject H_0, meaning there is a statistically significant difference between treatment groups. Therefore, this difference in means we observe between the samples is probably indicative of some underlying trend - rather than an instance of random chance.

8 Hypothesis Tests for Independence and Correlation

8.1 Chi-Square Test for Independence

The Chi-square test for independence is used to determine whether there is a statistically significant relationship between two categorical variables. For example, a researcher may want to evaluate whether an individual's level of physical activity is related to their dietary habits. The hypotheses for this test are:

H_0: The two categorical variables (e.g., physical activity level and dietary habits) are independent.
H_a: The two categorical variables are dependent.

Steps to Conduct a Chi-Square Test for Independence

1. **State the null and alternative hypotheses.**
2. **Choose the significance level** ($\alpha = 0.01, 0.05, 0.10$). This is the permitted probability of making a Type I error.
3. **Verify that all expected frequencies are at least 5.**
   This assumption ensures the validity of the Chi-square approximation. To check this:
   (a) Calculate the expected frequency for each cell using:
   $$e_{ij} = \frac{(\text{Row Total}) \times (\text{Column Total})}{\text{Sample Size}}$$
   (b) Examine each expected frequency in the table. If any value is less than 5, consider combining categories or using Fisher's Exact Test instead.
4. **Compute the test statistic** using:
   $$\chi^2_{df} = \sum\sum \frac{(o_{ij} - e_{ij})^2}{e_{ij}}$$
   where:
   - o_{ij} is the observed frequency for row i and column j,
   - e_{ij} is the expected frequency for row i and column j,
   - Degrees of freedom: $df = (r-1)(c-1)$.
5. **Find the *p*-value** using the Excel command below.
   ```
   =CHISQ.DIST.RT(test statistic, df)
   ```
6. **Make a decision:**
   - If p-value $< \alpha$, reject H_0.
   - If p-value $\geq \alpha$, do not reject H_0.

Example Problem

A researcher wants to determine whether an individual's decision to adopt a vegetarian diet is related to their frequency of exercise. A survey is conducted on 500 individuals, categorizing them by their diet choice (vegetarian or non-vegetarian) and their exercise frequency (low, moderate, high). The observed data is presented in the contingency table below:

	Low	Moderate	High	Row Total
Vegetarian	40	85	75	200
Non-Vegetarian	90	110	100	300
Column Total	130	195	175	500

Step 1: State the Hypotheses

H_0: Diet choice and exercise frequency are independent.
H_a: Diet choice and exercise frequency are dependent.

Step 2: Choose Significance Level

We choose $\alpha = 0.05$.

Step 3: Compute Expected Frequencies

Expected frequencies are calculated using:

$$e_{ij} = \frac{(\text{Row Total}) \times (\text{Column Total})}{\text{Sample Size}}$$

Applying this formula, the expected frequency table is:

	Low	Moderate	High	Row Total
Vegetarian	$\frac{(200)(130)}{500} = 52$	$\frac{(200)(195)}{500} = 78$	$\frac{(200)(175)}{500} = 70$	200
Non-Vegetarian	$\frac{(300)(130)}{500} = 78$	$\frac{(300)(195)}{500} = 117$	$\frac{(300)(175)}{500} = 105$	300
Column Total	130	195	175	500

Verification of the Assumption: Each expected frequency value is checked against the rule that all must be at least 5. Here, the smallest expected frequency is 52, which satisfies the assumption. If any expected frequency were less than 5, adjustments such as category combination or using Fisher's exact test would be necessary.

Step 4: Compute the Test Statistic

Using:

$$\chi^2_{df} = \sum \sum \frac{(o_{ij} - e_{ij})^2}{e_{ij}}$$

we calculate:

$$\begin{aligned}\chi^2 &= \frac{(40-52)^2}{52}+\frac{(85-78)^2}{78}+\frac{(75-70)^2}{70}+\\ &\frac{(90-78)^2}{78}+\frac{(110-117)^2}{117}+\frac{(100-105)^2}{105}\\ &=\frac{144}{52}+\frac{49}{78}+\frac{25}{70}+\frac{144}{78}+\frac{49}{117}+\frac{25}{105}\\ &=2.77+0.63+0.36+1.85+0.42+0.24\\ &=6.27\end{aligned}$$

Step 5: Find the p-value

Degrees of freedom: $df = (2-1)(3-1) = 2$. Using Excel:

```
=CHISQ.DIST.RT(6.27,2)
```

This gives a p-value of approximately 0.043.

Step 6: Decision

The p-value $= 0.043 < 0.05$, we reject H_0. This suggests that diet choice and exercise frequency are dependent.

Conclusion: The data provides evidence that an individual's diet choice is related to their exercise frequency.

8.2 Hypothesis Test for the Correlation Coefficient

The sample correlation coefficient r_{xy} measures the direction and strength of the linear relationship between two variables x and y. It is computed as:

$$r_{xy} = \frac{S_{xy}}{S_x S_y}$$

where:

- S_{xy} is the sample covariance,
- S_x and S_y are the sample standard deviations of x and y, respectively.

The value of r_{xy} ranges from -1 to 1:

- $r_{xy} = 1$ indicates a perfect positive linear relationship.
- $r_{xy} = -1$ indicates a perfect negative linear relationship.
- $r_{xy} = 0$ suggests no linear relationship.

Values closer to ± 1 signify a stronger linear relationship.

To determine whether the sample correlation represents the population correlation ρ_{xy}, we perform a hypothesis test.

Steps for Hypothesis Testing for the Population Correlation Coefficient

1. **State the Hypotheses:**
   - Two-tailed: $H_0 : \rho_{xy} = 0$ vs. $H_A : \rho_{xy} \neq 0$
   - Right-tailed: $H_0 : \rho_{xy} \leq 0$ vs. $H_A : \rho_{xy} > 0$
   - Left-tailed: $H_0 : \rho_{xy} \geq 0$ vs. $H_A : \rho_{xy} < 0$
2. **Select the significance level** ($\alpha = 0.01, 0.05, 0.10$). This is the permitted probability of a Type I error.
3. **Calculate the sample correlation coefficient in Excel:** `=CORREL(column_x, column_y)`
4. **Compute the test statistic:**

$$t_{df} = \frac{r_{xy}\sqrt{n-2}}{\sqrt{1-r_{xy}^2}}, \quad df = n - 2$$

5. **Find the *p*-value in Excel:**
   - Right-tailed: `=T.DIST.RT(test_statistic, df)`
   - Left-tailed: `=1-T.DIST.RT(test_statistic, df)`

- Two-tailed: `=2*T.DIST.RT(ABS(test_statistic), df)`

6. **Decision Rule:**

   - If $p < \alpha$, reject H_0.
   - If $p \geq \alpha$, do not reject H_0.

Example: Income and Education

Using the *Income Education* dataset in the LectureData.xlsx file, we test whether the correlation between income and education differs from zero at $\alpha = 0.05$.

Step 1: State Hypotheses

$$H_0 : \rho_{xy} = 0 \quad \text{vs.} \quad H_A : \rho_{xy} \neq 0$$

Step 2: Choose $\alpha = 0.05$

Step 3: Compute Sample Correlation

Using Excel: `=CORREL(A2:A27, B2:B27)` We obtain $r_{xy} = 0.8452$.

Step 4: Compute Test Statistic

$$\begin{aligned} t_{24} &= \frac{0.8452 \times \sqrt{24}}{\sqrt{1 - 0.8452^2}} \\ &= \frac{0.8452 \times 4.8990}{\sqrt{0.2857}} \\ &= \frac{4.1405}{0.5345} \\ &= 7.7465 \end{aligned}$$

Step 5: Compute *p*-value

For a two-tailed test: `=2*T.DIST.RT(ABS(7.7465), 24)` This returns $p = 5.5536E-$ .08.

Step 6: Decision

In this case, $p < \alpha$, we reject H_0.

Conclusion: At the 5% significance level, we conclude that income and education are significantly correlated.

9 Regressions Part 1

9.1 Simple and Multiple Linear Regression Model

Regression analysis is widely used in business, engineering, and the social sciences to evaluate how a response variable is influenced by explanatory variables. For example, the selling price of a house (response variable) can be predicted based on its size and location (explanatory variables). Regression models can be effective for understanding complex relationships that include multiple factors.

While correlation measures the strength of a linear relationship between two variables, regression extends this analysis to model the relationship between a response variable and one or more explanatory variables.

Deterministic versus Stochastic Relationships

- A **deterministic** relationship does not include any component of randomness. For example, momentum is always the product of mass and velocity. There are no other components in that relationship beyond mass and velocity.
- A **stochastic** relationship involves randomness or unmeasured factors. For instance, education affects income, but skill level—an unmeasured variable—also plays a role.

Simple Linear Regression

A **simple linear regression model** has one response variable and one explanatory variable:

$$y = \beta_0 + \beta_1 x + \epsilon$$

where:

- β_0 is the intercept (constant),
- β_1 is the coefficient of the explanatory variable,
- ϵ represents random error.

  In a deterministic relationship, $\epsilon = 0$.

If $\beta_1 > 0$, the relationship is positive; if $\beta_1 < 0$, the relationship is negative.

Example

Consider the regression model:

$$\text{Annual Income} = 3000 + 2000 \times (\text{Years of Education}) + \epsilon$$

- With 1 year of education: $5000 = 3000 + 2000(1)$

- With 2 years of education: $7000 = 3000 + 2000(2)$
- With 0 years of education: $3000 = 3000 + 2000(0)$

Each additional year of education increases income by \$2000.

Ordinary Least Squares (OLS) Estimation

The **Ordinary Least Squares (OLS)** method estimates β_0 and β_1 by minimizing the sum of squared errors:

$$e = y - \hat{y}$$

where y is the observed value and $\hat{y}$ is the predicted value.

The OLS estimates for β_0 and β_1 are:

$$\beta_1 = \frac{\sum(x_i - \bar{x})(y_i - \bar{y})}{\sum(x_i - \bar{x})^2}$$

$$\beta_0 = \bar{y} - b_1\bar{x}$$

In practice, software like Excel performs these calculations.

Excel Steps for Regression

1. Enable the **Analysis ToolPak** in Excel.
2. Open the dataset (e.g., Income and Education data).
3. Select **Data > Data Analysis > Regression**.
4. Specify the response variable (Income) as the Y range and the explanatory variable (Education) as the X range, including the first row in the selection.
5. Check the "Labels" box.
6. Click "ok"

The output should resemble the tables below:

SUMMARY OUTPUT

Regression statistics	
Multiple R	0.845
R Square	0.714
Adjusted R square	0.702
Standard error	12.125
Observations	26.000

ANOVA

	df	*SS*	*MS*	*F*	*Significance F*
Regression	1	8822.835669	8822.84	60.01	5.55229E-08
Residual	24	3528.548947	147.023		
Total	25	12351.38462			

	Coefficients	*Standard error*	*t Stat*	*P-value*	*Lower 95%*	*Upper 95%*	*Lower 95.0%*	*Upper 95.0%*
Intercept	-14.164	11.773	-1.203	0.241	-38.461	10.133	-38.461	10.133
Education (years)	5.734	0.740	7.747	0.000	4.206	7.262	4.206	7.262

Image 9.1: Regression Output

Interpreting the Output

The regression equation:

$$\text{Income} = -14.164 + 5.734 \times \text{Education}$$

- The coefficient 5.734 means one additional year of education is associated with an estimated \$5,734 increase in income.
- The intercept -14.164 suggests that when years of education is 0, the predicted income is \$-14,164.

Prediction Example

To predict income when education (years) is 16:

$$\text{Income} = -14.164 + 5.734(16) = 77.58$$

Thus, the estimated income for 16 years of education is \$77,580.

Multiple Linear Regression

A **multiple linear regression model** extends simple regression to include multiple explanatory variables:

$$y = \beta_0 + \beta_1 x_1 + \beta_2 x_2 + \beta_3 x_3 + \epsilon$$

Each coefficient represents the effect of that corresponding variable on the response variable while holding the other explanatory variables constant.

9.2 Regression with Dummy Variables

Regression models often include categorical variables, which must be converted into numerical values before analysis. A common approach is using **dummy variables**, which take values of 1 or 0 to represent different categories.

For example, in modeling professor salaries, numerical variables like experience (years) can be directly included in regression. However, categorical variables such as sex (male/female) or age group (below 60/60 and above) require dummy variables.

Dummy Variables in Regression

A **dummy variable** is defined as:

- $d = 1$ for one category,
- $d = 0$ for the other category.

For instance, if d_1 represents sex:

$$d_1 = \begin{cases} 1, & \text{if male} \\ 0, & \text{if female} \end{cases}$$

Similarly, for age:

$$d_2 = \begin{cases} 1, & \text{if age} \geq 60 \\ 0, & \text{if age} < 60 \end{cases}$$

Example: Professor Salaries

Given the regression model:

$$\hat{y} = 40.6060 + 1.1279x + 13.9240d_1 + 4.3428d_2$$

where:

- y = salary (in \$1000s),
- x = years of experience,
- d_1 = 1 if male, 0 if female,
- d_2 = 1 if age $\geq$ 60, 0 otherwise.

Predicting Salary

For a 50-year-old male professor with 10 years of experience ($d_1 = 1$, $d_2 = 0$, $x = 10$):

$$\begin{aligned} \hat{y} &= 40.6060 + 1.1279(10) + 13.9240(1) + 4.3428(0) \\ &= 40.6060 + 11.279 + 13.9240 \\ &= 65.809 \quad (\text{or } \$65,809) \end{aligned}$$

Regression Results

The estimated regression coefficients are shown below:

Variable	Coefficient	Standard error	p-value
Intercept (β_0)	40.6060	2.4501	0.000
Experience (β_1)	1.1279	0.1543	0.000
Male (β_2)	13.9240	2.8752	0.000
Age ≥ 60 (β_3)	4.3428	3.1204	0.162

Table 9.1: Regression results for professor salaries

Creating Dummy Variables in Excel

To create dummy variables in Excel using the `IF` function: =IF(logical test, value if true, value if false).

Avoiding the Dummy Variable Trap

For categorical variables, we must use $k - 1$ dummy variables to avoid multicollinearity (the **dummy variable trap**). For example, sex has two categories in this scenario (Male, Female), defined as:

$$d_1 = \begin{cases} 1, & \text{Male} \\ 0, & \text{Female} \end{cases}$$

Female is the **reference category**, meaning its effect is captured by the intercept of the regression. Instead of having two separate binary variables to define male and female, we have one binary variable with two categories.

9.3 Model Assumptions and Common Violations

Linear regression is a powerful tool for policy analysis, but its reliability depends on certain assumptions. If these assumptions are violated, the regression estimates become biased or inefficient, leading to misleading conclusions.

Key Assumptions of Ordinary Least Squares (OLS)

1. **Linearity**
   - **Definition:** The relationship between each explanatory variable and the response variable is linear.
   - **How to check:** Use scatterplots to visualize relationships. A roughly straight-line pattern suggests linearity; curved patterns indicate nonlinearity.
   - **If violated:** Coefficients become misleading. Consider transforming variables (e.g., logarithms) or choosing better predictors.
2. **Independence of Observations**
   - **Definition:** Each observation is assumed to be independent. For instance, if students collaborate while completing a test, their scores would no longer be independent.
   - **How to check:** Assess the data collection process. Time-series data inherently violates this assumption.
   - **If violated:** Standard errors are incorrect, leading to unreliable significance tests. Newey-West robust standard errors can be used for time-series data.
3. **No Omitted or Hidden Variables**
   - **Definition:** The model includes all relevant explanatory variables.
   - **How to check:** If adding a variable significantly alters coefficients, the model was likely missing a key variable.
   - **If violated:** The model suffers from omitted variable bias (OVB), leading to incorrect estimates. The solution is to collect more data and include the missing variables.
4. **Homoscedasticity (Constant Variance of Errors)**
   - **Definition:** The variance of the error term remains constant across all observations.
   - **How to check:** Plot residuals. A consistent spread around zero suggests homoscedasticity, while a funnel shape suggests heteroscedasticity.
   - **If violated:** Standard errors are incorrect, affecting confidence intervals and hypothesis tests. White's robust standard errors can correct for this.

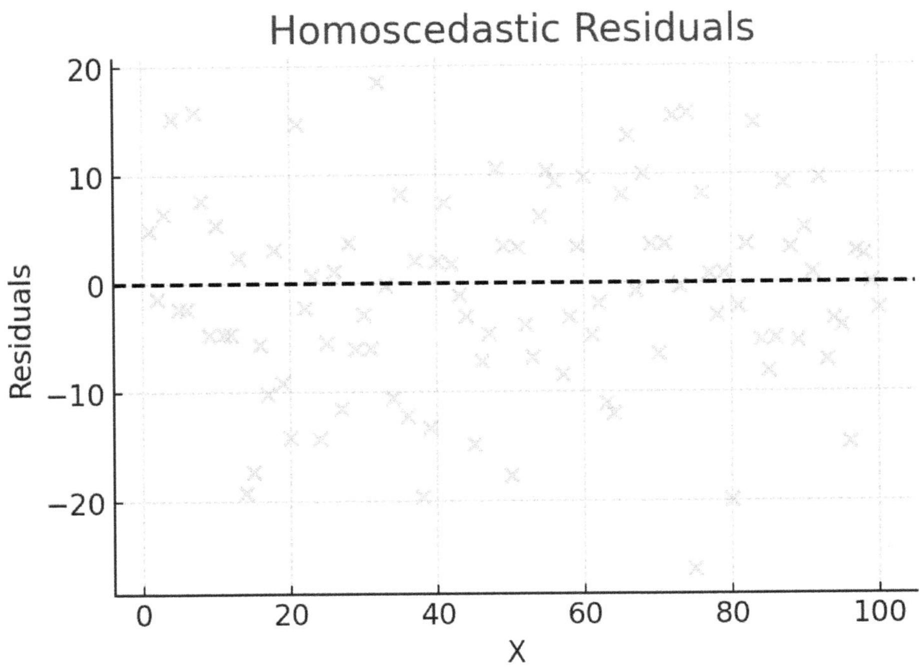

Figure 9.1: Example of homoscedastic residuals.

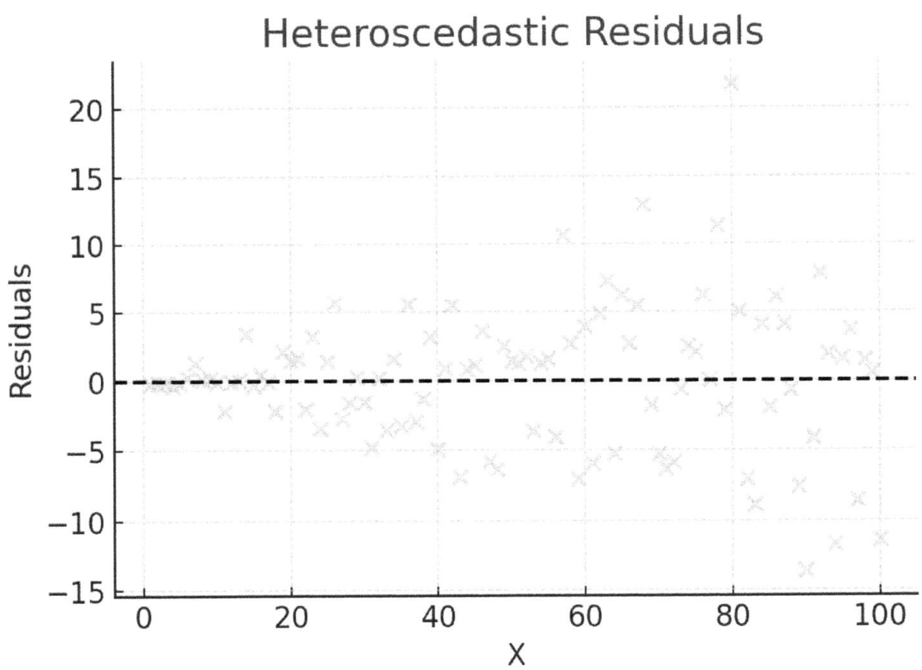

Figure 9.2: Example of heteroscedastic residuals.

5. **Normality of Residuals**
   - **Definition:** The distribution of residuals (errors) should follow a normal distribution.

- **How to check:** Use a density plot or histogram to ensure a bell-shaped distribution.
- **If violated:** The model may not be appropriate for small sample sizes. Transforming variables may help.

6. **No Multicollinearity (Explanatory Variables Should Not Be Correlated)**
    - **Definition:** Explanatory variables should not be highly correlated with each other.
    - **How to check:** Compute correlation coefficients in Excel (`=CORREL`). A correlation above 0.5 suggests multicollinearity.
    - **If violated:** Coefficients become unreliable. Consider removing one of the correlated variables.

7. **No Correlation Between Explanatory Variables and the Error Term (No Endogeneity)**
    - **Definition:** The error term should not be correlated with any explanatory variable.
    - **How to check:** There is no direct statistical test, but economic theory can suggest whether OVB is present.
    - **If violated:** Instrumental variables (IV) regression can be used as a solution.

8. **No Serial Correlation (Errors Should Not Be Correlated Across Observations)**
    - **Definition:** Error terms should not be systematically related over time.
    - **How to check:** Residual plots should show no discernible pattern. If errors display cycles or waves, there is serial correlation.
    - **If violated:** Newey West robust standard errors or time-series modeling techniques should be applied.

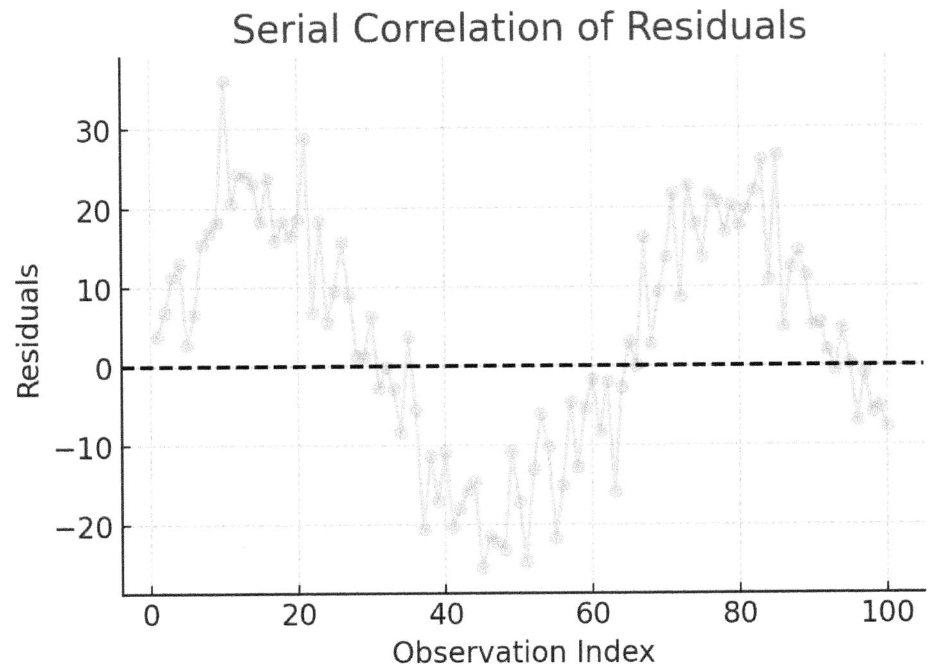

Figure 9.3: Example of serial correlation in residuals.

10 Regressions Part 2

10.1 Goodness-of-Fit Measures

Once we fit a linear regression model, we need to assess how well it explains the data. Three key measures help us evaluate the model's performance:

1. **Standard Error of the Estimate** (s_e) — Measures the average error in predictions.
2. **Coefficient of Determination** (R^2) — Measures the proportion of variance in the response variable explained by the model.
3. **Adjusted** R^2 — Adjusts R^2 for the number of explanatory variables to prevent overfitting.

1. Standard Error of the Estimate

The **standard error of the estimate** (s_e) quantifies the typical deviation of observed values from predicted values:

$$s_e = \sqrt{\frac{\sum(y_i - \hat{y}_i)^2}{n - k - 1}}$$

where:

- y_i = observed values,
- $\hat{y}_i$ = predicted values,
- n = number of observations,
- k = number of explanatory variables.

In practice, s_e is provided in regression output under **Regression Statistics**. In the income and education example, Excel reports $s_e = 12.125$.

SUMMARY OUTPUT

Regression statistics	
Multiple R	0.845
R Square	0.714
Adjusted R square	0.702
Standard error	12.125
Observations	26.000

ANOVA

	df	SS	MS	F	Significance F
Regression	1	8822.835669	8822.84	60.01	5.55229E-08
Residual	24	3528.548947	147.023		
Total	25	12351.38462			

	Coefficients	Standard error	t Stat	P-value	Lower 95%	Upper 95%	Lower 95.0%	Upper 95.0%
Intercept	-14.164	11.773	-1.203	0.241	-38.461	10.133	-38.461	10.133
Education (years)	5.734	0.740	7.747	0.000	4.206	7.262	4.206	7.262

Image 10.1: Regression Output

2. Coefficient of Determination (R^2)

R^2 represents the proportion of variation in the response variable explained by the model:

$$R^2 = 1 - \frac{SSE}{SST}$$

where:

- **SSE (Sum of Squares for Error)**: $\sum(y_i - \hat{y}_i)^2$ (unexplained variation).
- **SST (Total Sum of Squares)**: $\sum(y_i - \bar{y})^2$ (total variation).

A higher R^2 value (closer to 1) indicates better model fit. In the regression below, $R^2 = 0.714$, meaning the model explains 71.4% of the variation in income.

SUMMARY OUTPUT

Regression statistics	
Multiple R	0.845
R Square	0.714
Adjusted R square	0.702
Standard error	12.125
Observations	26.000

ANOVA

	df	SS	MS	F	Significance F
Regression	1	8822.835669	8822.84	60.01	5.55229E-08
Residual	24	3528.548947	147.023		
Total	25	12351.38462			

	Coefficients	Standard error	t Stat	P-value	Lower 95%	Upper 95%	Lower 95.0%	Upper 95.0%
Intercept	-14.164	11.773	-1.203	0.241	-38.461	10.133	-38.461	10.133
Education (years)	5.734	0.740	7.747	0.000	4.206	7.262	4.206	7.262

Image 10.2: Regression Output

3. Adjusted R^2

Since adding explanatory variables always increases R^2, the **adjusted** R^2 accounts for the number of variables:

$$\text{Adj. } R^2 = 1 - \frac{(1 - R^2)(n - 1)}{n - k - 1}$$

where:

- n = number of observations,
- k = number of explanatory variables.

Why use Adjusted R^2?

- Adjusted R^2 accounts for the number of explanatory variables in the regression model. More variables typically contributes to a higher R^2 value. Adjusting for this provides a more authentic R^2 value.
- Helps in model selection, preventing overfitting.

In Excel, adjusted R^2 is found in **Regression Statistics**. In our example, **Adj.** $R^2 = 0.702$.

SUMMARY OUTPUT

Regression statistics	
Multiple R	0.845
R Square	0.714
Adjusted R square	0.702
Standard error	12.125
Observations	26.000

ANOVA

	df	*SS*	*MS*	*F*	*Significance F*
Regression	1	8822.835669	8822.84	60.01	5.55229E-08
Residual	24	3528.548947	147.023		
Total	25	12351.38462			

	Coefficients	*Standard error*	*t Stat*	*P-value*	*Lower 95%*	*Upper 95%*	*Lower 95.0%*	*Upper 95.0%*
Intercept	-14.164	11.773	-1.203	0.241	-38.461	10.133	-38.461	10.133
Education (years)	5.734	0.740	7.747	0.000	4.206	7.262	4.206	7.262

Image 10.3: Regression Output

10.2 Test of Joint Significance

After constructing a linear regression model, we test whether the explanatory variables collectively contribute to predicting the response variable. This is known as the Test of Joint Significance, evaluated using an F-test.

The test examines whether all explanatory variable coefficients are jointly significant:

- Null hypothesis (H_0): All coefficients are zero ($\beta_1 = \beta_2 = \cdots = \beta_k = 0$).
- Alternative hypothesis (H_A): At least one coefficient is not zero.

We conduct a right-tailed F-test, where:

$$F_{(df_1, df_2)} = \frac{SSR/k}{SSE/(n-k-1)} = \frac{MSR}{MSE}$$

where:

- SSR = Regression sum of squares,
- SSE = Error sum of squares,
- k = Number of explanatory variables,
- n = Sample size,
- df_1 = Number of explanatory variables,
- $df_2 = n - k - 1$.

Example: Predicting Hospital Readmission Rates

Using healthcare data, we examine whether length of hospital stay (LOS) and patient age significantly predict hospital readmission within 30 days. The regression model is:

$$\text{Readmission Rate} = \beta_0 + \beta_1 \text{LOS} + \beta_2 \text{Age} + \epsilon$$

Regression Output:

Variable	Coefficient	p-value
Intercept (β_0)	15.23	0.002
Length of Stay (LOS) (β_1)	1.12	0.018
Age (β_2)	0.45	0.045

Table 10.1: Regression coefficients for hospital readmission

ANOVA Table for F-Test

The test statistic is derived from the ANOVA table:

Source	SS	df	MS	F
Regression (SSR)	120.45	2	60.23	28.68
Residual (SSE)	56.78	27	2.10	
Total (SST)	177.23	29		

Table 10.2: ANOVA table for readmission model

Know that the structure of a regression ANOVA table differs from that of a one-way ANOVA test.

Source	**df**	**SS**	**MS**	**F**	**p-value**
Regression	k	SSR	$MSR = \frac{SSR}{k}$	$F = \frac{MSR}{MSE}$	(from F)
Error (Residual)	$n-k-1$	SSE	$MSE = \frac{SSE}{n-k-1}$		
Total	$n-1$	$SST = SSR + SSE$			

Table 10.3: ANOVA Table for Regression

Hypothesis Test Steps

Step 1: State Hypotheses

$$H_0 : \beta_1 = \beta_2 = 0$$
$$H_A : \text{At least one } \beta \neq 0$$

Step 2: Choose Significance Level We set $\alpha = 0.05$.

Step 3: Compute Test Statistic From the ANOVA table:

$$F_{(2,27)} = \frac{SSR/k}{SSE/(n-k-1)} = \frac{60.23}{2.10} = 28.68$$

$$df_1 = 2$$

$$df_2 = 27$$

Step 4: Compute *P*-Value in Excel `=F.DIST.RT(28.68, 2, 27)` This returns $p = 2.0920E - 07$.

Step 5: Decision Rule Since $p < \alpha$ $(0.00 < 0.05)$, we reject H_0 and conclude that length of stay and age are jointly significant in predicting hospital readmission.

11 Regressions Part 3

11.1 Test of Individual Significance

The test of individual significance evaluates whether a specific explanatory variable significantly contributes to predicting the response variable. This is done using a t-test.

The test assesses whether a single coefficient is significantly different from zero:

- Null Hypothesis (H_0): $\beta_j = 0$ (the variable has no effect on the response variable).
- Alternative Hypothesis (H_A): $\beta_j \neq 0$ (the variable significantly affects the response variable).

We conduct a two-tailed t-test, where:

$$t_{df} = \frac{b_j - \beta_{j0}}{se(b_j)}$$

where:

- b_j = estimated coefficient from regression output,
- β_{j0} = hypothesized coefficient (typically 0),
- $se(b_j)$ = standard error of the coefficient,
- $df = n - k - 1$ (degrees of freedom, where n is the sample size and k is the number of explanatory variables).

Example: Predicting Hospital Readmission Rates

Using healthcare data, we examine whether LOS significantly predicts hospital readmission within 30 days. The regression model is:

$$\text{Readmission Rate} = \beta_0 + \beta_1\text{LOS} + \beta_2\text{Age} + \epsilon$$

Regression output:

Variable	Coefficient (b_j)	Standard error ($se(b_j)$)
Intercept (β_0)	15.23	2.05
Length of stay (LOS) (β_1)	1.12	0.23
Age (β_2)	0.45	0.21

Table 11.1: Regression coefficients for hospital readmission

Hypothesis Test Steps

Step 1: State Hypotheses

$$H_0 : \beta_1 = 0$$
$$H_A : \beta_1 \neq 0$$

Step 2: Choose Significance Level We set $\alpha = 0.05$.

Step 3: Compute Test Statistic Using values from the regression output:

- $b_j = 1.12$
- $\beta_{j0} = 0$
- $se(b_j) = 0.23$
- $df = 30 - 2 - 1 = 27$

$$t_{27} = \frac{1.12 - 0}{0.23} = \frac{1.12}{0.23} = 4.87$$

Step 4: Compute *p*-value in Excel `=2*T.DIST.RT(ABS(4.87), 27)` This returns $p = 4.32E - 05$ or 0.00004.

Step 5: Decision Rule We reject H_0 because $p < \alpha$ ($0.00004 < 0.05$). Therefore, we conclude that the length of hospital stay is a significant predictor of hospital readmission.

12 Regressions Part 4

12.1 Logistic Regressions

Linear regressions are used to analyze continuous outcome variables, like income, age, and stock prices. Logistic regressions analyze binary outcomes that represent a dichotomous relationship. For example, we could run a logistic regression that predicts whether a patient will develop sepsis in the hospital. The outcome (sepsis diagnosis) is a binary outcome.

Logistic Regression Model

$$log\frac{p}{1-p} = \beta_0 + \beta_1 x_1 + \beta_2 x_2 + ... + \beta_k x_k + \epsilon$$

Logit Distribution versus Normal Distribution

The defining difference between the logistic and linear regressions is the assumed underlying distribution of the outcome variable. The linear regression assumes that the outcome follows a normal distribution, whereas the logistic regression assumes the outcome adheres to the Bernoulli distribution. It is important to choose the right distribution because this dictates the accuracy of the interpretation of the p-values of the explanatory variables.

Interpreting the Output

The output interpretation of a logistic regression differs from that of a linear regression. Each coefficient of a logit regression represents the change in the log-odds of the outcome associated with a one-unit increase in that explanatory variable, holding the other variables constant. This can be difficult to provide a meaningful interpretation to a non-technical audience. Therefore, it's best to convert these coefficients to a more intuitive metric that can be easily understood. Two examples of preferred interpretations include the Marginal Effects and Odds Ratio.

1. **Marginal Effects**

   Interpretation: The marginal effects represent the change in the predicted probability for a one-unit increase in the explanatory variable.

   The marginal effects can be obtained by taking the derivative of the outcome variable with respect to the explanatory variable. Although this option is not easily attainable in Excel, it is so in Stata, R and SAS.

2. **Odds Ratio**

   Interpretation: The Odds Ratio represents the change in the odds of the response variable for a one-unit increase in the explanatory variable.

   The odds ratio can be obtained by calculating the exponent of the explanatory variable coefficient, using the $= EXP()$ command in Excel.

Example: Predicting Sepsis Using Age and Hospitalization History

We can use a logistic regression model to identify the relationship between age, hospitalization history and the presence of a sepsis diagnosis. Given the clinical work on this area, we know that older age groups with increased hospitalization episodes are at a higher risk of developing sepsis. However, this relationship is not well understood. Using the Sepsis dataset, we can develop a logistic regression, following the same steps in Excel as we did for the linear regression. The output is below.

SUMMARY OUTPUT

Regression Statistics	
Multiple R	0.931
R Square	0.866
Adjusted R square	0.864
Standard error	0.183
Observations	100.000

ANOVA

	df	*SS*	*MS*	*F*	*Significance F*
Regression	2	20.958	10.479	314.483	0.000
Residual	97	3.232	0.033		
Total	99	24.190			

	Coefficients	*Standard error*	*t Stat*	*P-value*	*Lower 95%*	*Upper 95%*	*Lower 95.0%*	*Upper 95.0%*
Intercept	-0.813	0.083	-9.820	0.000	-0.977	-0.649	-0.977	-0.649
Age	0.014	0.002	8.135	0.000	0.010	0.017	0.010	0.017
Hospitalization History	0.193	0.019	10.062	0.000	0.155	0.231	0.155	0.231

Image 12.1: Predicting Sepsis Using Age and Hospitalization History Regression Output

The positive coefficients for the age (0.014) and hospitalization history (0.193) variables demonstrate the positive effect these variables have on a sepsis diagnosis. Therefore, an increase in age or hospitalization history is associated with an increase in the log-odds of a positive sepsis diagnosis.

The exponentiated age coefficient (0.014) is 1.014, meaning that for each one-unit increase in age (one year), the odds of sepsis increase by about 1.4%, holding all other variables constant. The exponentiated hospitalization history coefficient (0.193) is 1.212, meaning that for each one-unit increase in hospitalization history (one previous healthcare episode), the odds of sepsis increase by about 21.2%, holding all other variables constant.

Credits

IMG 9.1: Copyright © by Microsoft.
Fig. 9.1: Generated using ChatGPT.
Fig. 9.2: Generated using ChatGPT.
Fig. 9.3: Generated using ChatGPT.
IMG 10.1: Copyright © by Microsoft.
IMG 10.2: Copyright © by Microsoft.
IMG 10.3: Copyright © by Microsoft.
IMG 12.1: Copyright © by Microsoft.

www.ingramcontent.com/pod-product-compliance
Ingram Content Group UK Ltd.
Pitfield, Milton Keynes, MK11 3LW, UK
UKHW051241300926
15471UKWH00015B/1485

9 798823 386968